Studying Human Behavior

Studying Human Behavior

How Scientists Investigate Aggression and Sexuality

Helen E. Longino

The University of Chicago Press • Chicago and London

Helen E. Longino is chair and the Clarence Irving Lewis Professor in the Department of Philosophy at Stanford University. She is the author of *Science as Social Knowledge* and *The Fate of Knowledge*.

The University of Chicago Press, Chicago 60637
The University of Chicago Press, Ltd., London
© 2013 by The University of Chicago
All rights reserved. Published 2013.
Printed in the United States of America

22 21 20 19 18 17 16 15 14 13   1 2 3 4 5

ISBN-13: 978-0-226-49287-2 (cloth)
ISBN-13: 978-0-226-49288-9 (paper)
ISBN-13: 978-0-226-92182-2 (e-book)
ISBN-10: 0-226-49287-7 (cloth)
ISBN-10: 0-226-49288-5 (paper)
ISBN-10: 0-226-92182-4 (e-book)

Library of Congress Cataloging-in-Publication Data

Longino, Helen E.
    Studying human behavior : how scientists investigate aggression and sexuality / Helen E. Longino.
        pages ;   cm
    Includes bibliographical references and index.
    ISBN 978-0-226-49287-2 (cloth : alkaline paper)
    ISBN 0-226-49287-7 (cloth : alkaline paper)
    ISBN 978-0-226-49288-9 (paperback : alkaline paper)
    ISBN 0-226-49288-5 (paperback : alkaline paper)
    ISBN 978-0-226-92182-2 (e-book)
    ISBN 0-226-92182-4 (e-book)
    1. Human behavior—Research.   2. Behaviorism (Psychology).
    3. Aggresssiveness—Research.   4. Sex—Research.   I. Title.
    BF199.L58   2012
    155.2'32—dc23
                                                            2012017786

♾ This paper meets the requirements of ANSI/NISO Z39.48-1992
(Permanence of Paper).

For Valerie

Contents

Acknowledgments

This book began life as two chapters excised from my 2001 book *The Fate of Knowledge*. As I completed that manuscript, it became clear that the sciences of human behavior I was using to exemplify scientific plurality required far more space than I could allot to them. My observations needed to be articulated in the context of more pressing questions raised by the very aim of understanding human behavior. At the time those included the conceptualization of behavior and the complicated relationships among the sciences, culture, and policy. These matters demanded treatment on their own merits.

As I have worked my way through this investigation, I have used a social epistemological methodology to understand the differences and similarities among the several approaches, focused on characterizing what the different approaches are dedicated to explaining, and followed the circulation of assumptions and values through multiple technical and cultural contexts. The sciences of behavior have recently been advanced by rapid changes in genetics and neurobiology, as well as in the psychological and social sciences. Nevertheless, the persistence of certain themes across the decades invites philosophical inquiry into the kind of knowledge these sciences provide about human behavior and how it matters.

•

I've been fortunate in being given opportunities to present parts of this work at national and international conferences and universities. The discussions on those occasions introduced me to new sources and helped me clarify my ideas. Journal and anthology editors who published early versions of parts of chapters both encouraged me to persist in the project and pushed me to think more clearly about the issues it encompassed. Lisa Lloyd and Ken Waters read early versions of chapters and talked with me at length and on many occasions. The comments of anonymous referees who read the manuscript were invaluable, leading me to restructure the entire book. Kenneth Kendler read a large portion of the draft and generously talked with me about issues in the research. Valerie Miner also offered excellent advice on various portions of the manuscript. Colleagues at Stanford University's Center for the Integration of Research on Genetics and Ethics and Clayman Institute for Gender Research offered helpful feedback, as did participants in reading groups at the University of Washington and Waterloo University, as well as the students in the fall 2010 and 2011 offerings of my course "Philosophy, Biology, and Behavior." Finally, Karen Darling's enthusiasm for the project sustained me in the final push to completion. Joel Score's smart and sensitive copyediting made the book more readable. I am deeply grateful to everyone who contributed to the shaping of this book. Responsibility for its remaining errors, of course, rests with me.

.

I am further grateful to the University of Minnesota for grants and leave that enabled me to devote time to the project and to employ research assistants to comb through large bodies of literature; to the National Science Foundation, which in 1998 and again in 2004 gave me grants (SBR 9730188 and SBR 0424103) that provided time for research; and to Stanford University for research support, especially for the engagement of research assistants who helped with the citation analyses of chapter 10. Over the years many research assistants have contributed to the project, including Steven Fifield, Kristen Houlton, Yeonbo Jeong, Sam Bullington, Cale Basaraba, Mario Silva, Whitney Sundby, and Elizabeth Lunstrum. And the friends of Bill W. continue their support.

.

Portions of chapters appeared, in different form, in several venues.
    I am grateful to Elsevier, Ltd., for permission to incorporate portions of my article "What Do We Measure When We Measure Aggression?" (*Studies in History and Philosophy of Science,* fall 2001) in chapters 1 and 9.
    I am grateful to The University of Minnesota Press for permission to use portions of "Theoretical Pluralism and the Scientific Study of Behavior," first pub-

lished in 2006 in *Scientific Pluralism*, edited by Steven Kellert, Helen Longino, and C.K. Waters (volume 19 in the series Minnesota Studies in Philosophy of Science), in chapter 8.

In addition, parts of chapter 8 draw on arguments presented in earlier form in my "Evidence in the Sciences of Behavior," in *Evidence*, edited by Peter Achinstein (Johns Hopkins University Press, 2005), and "Knowledge for What?," in *Philosophy of Behavioral Science*, edited by Kathryn Plaisance and Thomas Reydon (Kluwer, 2011). Parts of chapters 9 and 10 were prefigured in "Behavior as Affliction: Framing Assumptions in Behavior Genetics," in *Mutating Concepts, Evolving Disciplines: Genetics, Medicine, and Society*, edited by Rachel Ankeny and Lisa Parker (Kluwer, 2002).

•

Scientific inquiry and knowledge are social, I have argued. So is philosophical inquiry. I am grateful to all who have entered into this dialectical adventure with me.

1

Introduction

Efforts to understand human behavior take many forms. Philosophers have argued about the extent and nature of will: Is it free, or are our choices the outcomes of natural events beyond our consciousness? Can our common-sense understandings of human action and behavior be reconciled with our understandings of how the physical world works? Novelists and playwrights have examined the relations of character, situation, action, and outcomes. Scientists have investigated evolutionary history to find out about the emergence of distinctively human forms of behavior and investigated the brain and nervous system to understand the neural underpinnings of action. They have studied the parental and pedagogical environment in which children acquire habits and dispositions to understand the role of social environment in the development of individuals, studied intrafamilial similarities to understand the role of shared genes in engendering dispositions, and proposed various ways of integrating the understandings gleaned from such studies. What is entertained in any of these arenas has a bearing on the plausibility and implausibility of ideas entertained in any of the others, but, for the most part, the philosophical, creative, and scientific inquiries carry on quite independently. In spite of this independence, when it comes to designing institutions, selecting policies, and adopting therapeutic practices, we place priority on being informed, where possible, by scientific research.

Given the variety of approaches to studying human behavior scientifically, by which should we seek to be informed? How should we even go about answering this question? And what do different ways of answering it assume about the nature of scientific knowledge? Just as there is an age-old question about the nature of the will, so is there a distinct, but equally old, question about human differences and similarities: are we humans products of nature or of nurture? There are more refined ways of asking this question, and we will encounter them in this book. But the question, crude as it is, animates the public and media interest in scientific research on behavior. Its salience also shapes the overall pattern of relations among different research approaches to the study of human behavior.

In this book, I examine a set of research approaches that are in one way or another engaged with the debate to understand their epistemological structure (investigative methods, assumptions, basic concepts), the kinds of knowledge they provide, and the pragmatic aims they can be seen to advance. My aim, unlike that of many recent writers on this subject, is not to enter the nature/nurture debate. Focusing on that dichotomy and on research that seems to favor one side or the other distracts both researchers and the public from attending to other questions about human behavior that may be just as valuable in addressing the pragmatic concerns we look to the sciences to help us resolve.

My examination is informed by the social epistemological account of scientific inquiry for which I have argued in earlier books.[1] This philosophical account treats scientific knowledge as an outcome of critical interaction among different perspectives as well as of empirical engagement with a given subject matter. On this view, philosophical appraisal of research should encompass the full range of research approaches pursued in order to evaluate any one of them. This enables the analyst to see the full range of data that are collected, the range of hypotheses under consideration, and the assumptions in light of which data acquire evidential relevance to those hypotheses. While in some cases, one approach may turn out to be more empirically successful than the alternatives, in other cases, inquiry may sustain a multiplicity of viable approaches, each generating some knowledge. The question becomes not which is the best or better, but what each contributes—both in terms of positive results and in terms of critical perspectives on the others—to our overall understanding of a given phenomenon. This openness to plurality of approaches constitutes a philosophical pluralism.

The subject matter of my study is empirical research on behavior. I ask what can be known about human behavior through empirical investigation. I examine classical and molecular genetic approaches, social-environmental approaches, neuroanatomical and neurophysiological approaches, several approaches that

---

1. Longino 1990, 2001b.

attempt to integrate the factors studied in the aforementioned approaches, and, finally, what I call population or human-ecological approaches. Empirical research must have an object that can be observed and measured. Behavior is too vague a notion to constitute the subject of a study. Empirical studies look at particular behaviors, operationalized in specific ways that permit measurement. I have analyzed research on two families of behavior: aggressive behavior and sexual behavior. In part this is a matter of convenience and history: their study was the partial subject of an earlier book and I was consequently familiar with some of the issues.[2] But these behaviors are also subjects of intense interest, both scientific and social, with the result that there are many studies representative of the various research approaches, studies that have generated sufficient material to make comparative analysis feasible. Furthermore, in the course of this project, I have become convinced that understanding the structure of research on these behaviors, in particular, holds very important lessons for our mode of reliance on scientific research in the design of policy and institutions.

Political Contexts

Worries about crime and violence. The politically charged character of the nature-nurture debate is evident in a controversy that overtook what started out as an attempt to assess genetic research on aggression. Rates of violent crime in the United States rose steadily through the late 1970s and 1980s, peaking in the early 1990s and beginning a slow decline in the mid- to late 1990s.[3] In 1992 a conference entitled "Genetic Factors and Crime" was scheduled to take place on the campus of the University of Maryland with funding from the United States National Institutes of Health.[4] Plans for the conference attracted the criticism of Peter Breggin, a psychiatrist and vocal opponent of biological psychiatry and the use of psychotropic medications. His charge of racism in the conceptual and theoretical foundations of the conference got the United States Congressional Black Caucus involved. The Caucus exerted pressure on the National Institutes of Health, which ultimately withdrew funding for the conference. Conference organizers responded by threatening a lawsuit against the NIH. The ensuing negotiations resulted in the rescheduling of the conference. It was held three years

---

2. Longino 1990.

3. http://www.bjs.gov/index.cfm.

4. See Holden (1992), Palca (1992), Stone (1992), Charles (1992), Touchette (1995), and Roush (1995) for accounts of the original conference cancellation and of the conference that did finally occur.

later, in 1995, in a less prominent location—a university-owned conference center on Maryland's eastern shore—and with a broader range of participants, including academic critics of biological research on social behavior and researchers working from other perspectives. Uninvited "political" critics also showed up and voiced their concerns both to conference participants and to journalists covering the affair.

The main concerns of the protesters were possible racist uses of the work or racist agendas underlying the funding of the research. There were at least two reasons for this concern. The first resides in the social milieu of the conference and of much of the research being discussed. Even now, researchers in this field include few African Americans.[5] Americans not of African origin still carry stereotypes of African American men as bestial and as prone to criminality.[6] Unemployment among African Americans is twice that of white Americans, and periods of unemployment for African American individuals last longer.[7] In the United States, men of African origin are represented in the prison population at a far higher rate than in the general population.[8] In the late 1980s and early 1990s about one-third of African American men between the ages of twenty and twenty-nine were either in prison, on probation, or on parole at any given time.[9] Now, there are many ways to account for the disparity between African American and white arrest, conviction, and incarceration rates. Racial bias in the criminal justice system is one such factor. The overrepresentation of African Americans among the urban poor is another. In spite of significant formal gains since the civil rights movement of the 1950s and 1960s, and the election of an African American to the presidency of the United States in 2008, much remains to be done to achieve full racial equality, and the psychological residue of formerly legal racist bias persists.

The worry about genetic research in this context is that, to the extent that it located the springs of violence within the individual, thus making it associable with other individual properties such as race, it would support treating the disparity as an outcome of natural differences. This would direct attention away

5. For recent data on African Americans in the sciences, see National Science Foundation 2009.
6. Goff et al. (2008) detail social psychological studies demonstrating such continued stereotyping.
7. Lee and Morse 2004.
8. It cannot be too often said that the majority of victims of crime are also African American.
9. Stern 1997, 50–51; Alexander (2010) suggests little change in these figures. Pettit (2009) and Pettit and Western (2004) document the effects of disproportionate incarceration of African Americans on African American communities.

from social factors and, hence, weaken support for alleviating poverty and eliminating racism in the judicial system. These concerns are similar to those that were generated by earlier research attempting to correlate race and intelligence. There the issue was whether the differential performance of members of different racial groups on standardized tests was a function of innate, genetically determined abilities, rather than a function of differential access to education or of cultural bias in the tests themselves or in the circumstances of test-taking.[10] The critics of the violence conference recognized that the very fact that the genetic question is taken seriously by scientific researchers confers legitimacy on this view, whatever the eventual outcome of the research.[11] This is the case regardless of the explicit motives of the researchers. That such uses might actually be part of the agenda was made plausible by ill-considered remarks by the then director of the Alcohol, Drug Abuse, and Mental Health Administration, Frederick Goodwin, likening inner cities to jungles. He was undoubtedly trying to underscore the urgency of the issue of crime, but the suggestion of aspects of social Darwinism was not lost on critics of the conference.

A secondary concern derives from the more restricted context of psychoneuropharmacology. Persons opposed to what they perceive as psychiatric overreaching are concerned about the abuse of pharmacological intervention on those identified as "at risk" for engaging in antisocial behavior. They are concerned, that is, about a pharmacological normalization or pacification of the population. The primary and secondary concerns interact in political activists' expressed worry that African American children will be medicated on a wide scale as a consequence of this research. One might be tempted to dismiss this concern as unfounded or outdated. That this fear has had such currency in many sectors of African American society, however, is a sign of the deep mistrust with which African Americans regard the governing establishment in the United States and, as such, is not to be lightly dismissed.[12]

Debates about sexual orientation. The nature-nurture debate swirls too around sexual orientation, as controversy about homosexual clergy roils established religious denominations like the Lutherans and Episcopalians, as some countries

---

10. Jensen 1969; Steele 1997; Steele and Aronson 2004. The literature here is vast, as it is also for questions about the basis of observed gender differences in behavior and measures of intellectual performance.

11. See Lewontin 1991.

12. For a wide-ranging critique of many potential social implications of genetic research, see Hubbard and Wald 1993; also Kaplan 2000.

and local jurisdictions extend marriage rights to gay couples while others act to bar any such change, and as some continue to expel those who are open about their sexuality.[13] Political operatives in the United States have notoriously used antigay sentiments to motivate socially conservative voter turnout. Until the late 1960s there was an unspoken compact in the United States concerning issues of sexuality. Monogamous heterosexual marriage was the norm. Exceptions were taken as evidence of immorality or psychological deviance and were not spoken of in polite company. Psychiatric classification of homosexuality as a mental disorder and the criminal status of homosexual sex acts consigned men and women in same-sex sexual relationships to social invisibility, leaving them vulnerable to blackmail, job loss, and thuggery. The *Diagnostic and Statistical Manual of Psychiatric Disorders* removed homosexuality from its list in 1973, and the United States Supreme Court declared laws against sodomy unconstitutional in 2003. The compact of silence broken, modern industrial societies have become more open to and tolerant of same-sex sexual relationships, while antigay voices have become increasingly strident.

The lines of public debate concerning the causes of homoerotic and homosexual orientation are complicated. Whereas most African Americans see as injurious research purporting to show a genetic basis for group differences in intelligence test performance or, potentially, aggression, gay activists differ in their appraisal of genetic and biological research into the etiology of sexual orientation. Researcher and activist Dean Hamer claims that acceptance of a genetic basis or cause for homosexuality will dissuade people from homophobic attitudes and make them more supportive of extending civil rights to gay people.[14] He sees genetics and choice as the two alternative causal narratives: a choice can be punished or morally condemned, while a genetic outcome cannot be and hence must be tolerated. Other activists counter that the scientific research does not support a general thesis about the biological basis of sexual orientation and that identifying such a basis would do little or nothing to moderate homophobic tendencies.[15] Belief in such a basis would simply encourage research into biological "cures" for homosexuality. And were a gene or gene complex found to be associated with homosexuality, parents might opt for fetal testing that would

---

13. The "don't ask, don't tell" policy providing for dishonorable discharge of identified gay men and lesbians from the United States armed services was repealed as this manuscript went to the publisher.

14. Hamer and Copeland 1994; Hamer 2006.

15. Schüklenk 1993.

enable them to abort fetuses carrying it, just as parents in countries with a cultural preference for offspring of one or the other sex now engage in prenatal sex identification and selective abortion.[16]

•

Given these social contexts, one might think it impossible to conduct impartial research into the nature and causes of behaviors covered by the rubrics of aggression and sexual orientation. But any perusal of journals of human genetics, behavioral science, psychiatry, social science, aggression, or sexuality reveals an ocean of research articles, meta-analyses, and review articles concerning such behaviors, along with theoretical and programmatic statements defending the various frameworks under whose aegis the research is conducted. This book is an effort to sort through a significant subset of this literature. My aim is to delineate the evidential and argumentative structure of empirical research on human behavior that either employs one or another biological approach or is claimed to present an alternative to such approaches. Public discussion of this research usually frames the issue as a substantive question about the causes of behavior or of differences in behavior. Among researchers the debates are understood somewhat differently: as debates about what subfields (if any) of the biological and related sciences are best adapted, most appropriate, for the study of human behavior, or, to put it another way, what research tools are most useful for such studies. Positions regarding methods, however, involve assumptions about the substantive matters. The latter, then, cannot be dismissed as part of a public (mis)understanding of scientific research.

I argue for and elaborate three principal theses, one epistemological, regarding the character of knowledge generated in this research, one ontological, regarding the object of knowledge, and one social, regarding the differential uptake and diffusion of knowledge. Epistemologically, the relation among the different research approaches is neither competitive, reductionist, nor additive but is best understood in a pluralist framework. Each approach offers partial knowledge

---

16. Greenberg and Bailey 2001; West 2001. As will be seen, even where research does suggest an association, it is far from the 95% or greater association between chromosomal sex and anatomical/physiological sex. While selection procedures could in principle be extended to fetuses of either sex, and some parents might be trying to maximize their chances of a female child, most contexts in which sex selection is routinely practiced are male-preferring, reflecting cultural preferences for males. See Chan, Blyth, and Chan, 2009; Booth, Verma, and Beri 1994; Chu 2001; Sen 1990.

that need not be congruent or commensurable with knowledge produced in another approach, even when they have overlapping subject matter. Ontologically, there is no single concept of behavior. Furthermore, the operationalization of particular behaviors reflects folk psychological and moral values or political concerns as much as objective properties of the phenomena. Socially, there is not enough epistemologically productive interaction among these approaches, and the patterns of research uptake distort the perception of what scientific investigation can reveal.

Preliminaries

Studying Behavior

A given behavior can be of either primary or secondary interest to researchers. When it is the primary subject of investigation the empirical question may concern the frequency, distribution, or etiology of a given behavior or behavioral disposition.[17] Conceptual questions have to do with what behavior is—bodily movement, intentional action, situated interaction. A mixed conceptual and empirical issue concerns the appropriate classification and distinctions to introduce regarding a particular behavior: how should it be operationalized, how distinguished from other behaviors, to which other behaviors related? Additionally, there are methodological issues: how best to study a given behavior understood as a phenomenon of a certain kind. Once these preliminaries are settled, the substantive empirical questions concerning frequency, distribution, and etiology can be addressed.

Behavior is a secondary subject of investigation in studies aimed at understanding some factor thought to be involved in the production or performance of some kind of behavior. For example, researchers studying the behavioral effects of manipulating serotonin uptake or levels of gonadal hormones may be primarily interested in understanding the scope of effects of a physiologically active substance. While popular reporting may suggest otherwise, behavior is, in these cases, a tool—a clue to the activity of that substance in the organism's ner-

---

17. Much research on behavior is addressed not to the causal factors involved in producing particular instances or tokens of a behavior but to the factors involved in the inculcation of dispositions to behave in certain ways. This is especially so in research focused on behavior as individual movement or action, as distinct from interaction.

vous system—and not itself the subject. Similarly, researchers may be interested in the cascade of effects caused by a given allelic mutation. In such cases, the research problem is formulated around the substance or gene, not around behavior per se. Measuring a change in behavior is important to identifying more proximate organic effects, but the behavior itself is of less interest than the gene or substance. While this may be true of the internal epistemological structure of the research, pragmatically the choice of behavioral effect to be studied can be very important in terms of attracting funding, establishing a model that will be used in other research programs, attracting positive public interest—business a sociologist of science might call enrolling allies. Such considerations can reach into the work to pull it in one direction rather than another, asserting exigencies that the internal structure must rearrange itself to satisfy.[18]

The nature-nurture debate is one of the most notorious and familiar of these boundary-crossing issues and is unavoidable in discussions of behavior. The debate has political versions: Are socially important differences between persons inherent in individuals or created in them by their society or upbringing? What can or should be done to minimize socially important differences between individuals? And it has theoretical versions: To what factors should one look to explain behavior? What factors are most important in the causation of behavior or behavioral dispositions? Because it is so pervasive an issue, likely to provoke immediate visceral responses, it is important to explain the ways in which it is and is not involved in the research I will be describing.

At least since Plato and Aristotle, thinkers have differed on the degree to which socially important human traits are malleable. Some regard treating behavior as fixed by nature as politically progressive, while others regard it as politically reactionary. Philosophically, the dichotomy has been articulated within materialist, idealist, monist, dualist, and compatibilist metaphysical frameworks. In modern philosophy, the metaphysical issues have changed—the relation between the physical and the mental having become one of the central conundrums—but the nature-nurture debate persists.[19]

---

18. One of the most effective demonstrations of this interaction is Knorr-Cetina 1981.

19. Plato and Aristotle themselves differed in their overall metaphysical views, e.g., regarding the nature of universals, in their political views regarding the source of political authority, and in their conceptions of human nature. For example, Plato held that at least some gender differences were the product of training (and could be changed by different training), while Aristotle held that the sexes were different by nature, women being incomplete or lesser humans. Gendered properties for Aristotle were not mutable but essential aspects of human beings.

The resilient quality of the nature-nurture dichotomy suggests that it is generated by something deeper than metaphysical or empirical issues, or at least by something other than their most familiar versions. My concern with the debate is the ways it shapes both scientific research on behavior and public understanding of what that research can or should tell us. All behavioral research takes place against a background of ongoing debate in many discursive contexts about the relative influence of nature and nurture on the development of individuals in general and on any given behavior in particular. Outside the communities studying behavior, most people who take any interest in such issues tend to argue from one or the other side of the dichotomy. Advocates of different research approaches, by contrast, all acknowledge the roles of both nature and nurture (or genes and environment) as far as their own research is concerned, and most deplore the dichotomy of nature and nurture. Yet even among researchers, the dichotomy exerts a polarizing influence, as advocates of rival approaches are accused of creating and then falling onto one or the other horn of the dilemma. My aim, therefore, to the extent I am concerned with the nature-nurture question, is neither fully to settle nor to deconstruct the debate. It is instead to understand how persistent interest in the dichotomy interacts with other questions and assumptions in shaping behavioral science research.

The nature-nurture debate is sometimes, because of ideas about which is more amenable to intervention and control, represented as equivalent to the determinism-antideterminism debate. Nurture is often thought to be more malleable than nature, while determinism is identified with imperviousness to human intervention.[20] Opponents of biologically oriented approaches deplore genetic or biological determinism, but if the thought is that nonbiological, nurture approaches are better aligned with the free-will side of the debate, it is misguided. Both poles of the nature-nurture debate are equally deterministic (or nondeterministic)—the issue is *what* does the determining: biologically innate factors, social-environmental factors, or some complex of these. And given that all sides concede that both biology and environment are important, the debate is more precisely about two questions: (1) the relative strength of the influence of each and (2) the best methods for studying the determinants of behavior.

---

20. This depends, of course, on one's theory of social dynamics as well as on one's view of biological stability and change. Some theorists regard large-scale social dynamics as inflexibly law-governed and as having permanent formative effects on the individuals who participate in those processes. Others see them as far too complex either to have deterministic effects on individuals or to be susceptible to manipulation. In our brave new world of pharmacological mood and behavior management, nature may come to be perceived as the more malleable of the forces.

The Stakes—Real and Perceived

Both areas of research to be analyzed here engage issues relevant to highly sensitive aspects of social order. Research on aggression is intended to show us something about violence and crime. Studies of sexual orientation may show us something both about reproduction and about patterns of intimacy. In addition, the biological work in which behavior plays a secondary role is relevant to understanding gene action and activation, neurotransmitter and hormone function, and physiological dimensions of behavior. At stake in these debates, then, are methodologies for studying social/behavioral phenomena and the biology of human and other animals, as well as conceptions of the good life and the good society. This means that many different constituencies have stakes in the conduct and outcomes of this work: the scientists whose research is valued or devalued, used or neglected; politicians concerned with maintaining social order; communities wracked by violence or stereotyped as criminally violent; gay and bisexual men and women rejecting the label of deviant and demanding equal treatment under the law; members of the general public who, whether they concern themselves with this work or not, fund much of it through taxpayer-supported agencies and will be directly or indirectly affected by the conceptual and pragmatic directions the work takes.

As the story of the genetics and crime conference illustrates, not everyone has the same kind of access to knowledge production. Researchers working within an accepted approach and spokespersons from the research community arguing on behalf of such an approach have the greatest access—thus, in the original plan, the conference was limited to individuals working on programs connected with genetics. The conference that eventually took place, after the initial controversy, was expanded to include researchers from different research communities and traditions: biologists engaged in physiological rather than genetic research, biologists opposed to behavioral genetic research, and sociologically oriented criminologists, as well as historians and philosophers. While they would not have been included without the political protest that led to cancellation of the original conference, as academics they still represent the acceptable face of criticism. Political protesters, mostly from left-wing organizations, who saw themselves as representing communities potentially affected by the work were still excluded and felt compelled to force their way into the conference and "take it over."[21] Gay men and lesbians concerned over research on possible biological

---

21. Although they did succeed in getting a hearing and some researchers reported a shift in their way of thinking as a result, it would be a stretch to suggest that the protesters took it over. Some observers were left

bases or correlates of sexual orientation are better represented in the academy, and indeed, among the researchers, than are the urban poor and African Americans who may be (or feel themselves to be) threatened by the research on violence. The reception of this work has been mixed among gay men and lesbians, who have themselves organized conferences and forums to debate the theoretical and the social dimensions of the new research.

In addition to asking who can speak, one might ask, in Latourian fashion, *what* can speak, that is, what are the nonhuman participants "recruited" into these studies. These range from genes and genetic markers, neurotransmitters and their antagonists, hormones, metabolic states, and the instruments used to insert, manipulate, and identify these biological agents, to behavioral correlations, distributions, and the instruments used to identify and measure them (questionnaires, interview schedules, diagnostic categories, statistical methods, etc.), and, of course, the experimental animals from rats to drosophila that stand behind the human studies. These heterogeneous elements are assembled by researchers to construct relevant data. Contrary to the impression created by some articulations of actor-network theory, these factors are not independent, self-constituted players/participants but are assigned meaning within the different approaches in which they figure. Their significance must be established; it is not a given. That a particular individual has a specific resting heart rate under stipulated circumstances, or responds to a serotonin antagonist in a particular way, that there are particular ways a rat responds to an intruder or a mutant fruit fly interacts with another fruit fly, or that a given genetic marker is associated with a phenotypic trait in some population, are facts independent of what a researcher expects or desires. What is made of such data, how they are identified and represented, to what kinds of question they are deemed relevant, is a matter, however, of the assumptions and framework within which research is conducted, of the networks of facts that can be made to cohere sufficiently to support a way of going about inquiry.[22]

---

thinking that "the public" simply doesn't understand the relevance of scientific inquiry to the policy issues about which the protestors were concerned. See Jones 2006.

22. Terms like "stipulated circumstances," "in a particular way," and "associated" are not given by nature but constitute part of the conceptual framework within which facts are extruded (extracted as isolable and exportable from the dense and complex manifold of experience). For a fascinating approach to "facts," see the work of the Facts research group in the Economic History department of the London School of Economics: http://www.lse.ac.uk/collections/economicHistory/Research/facts/Default.htm.

Plan of the Book

The chapters that follow focus on research on two kinds of behavior—aggression and sexual behavior. Just as one might question the possibility of dispassionate research about the phenomena themselves, one might question using this body of research to support any conclusions about scientific inquiry into human behavior. On the other hand, these are central behaviors involved in social dispersion and cohesion, behaviors we need to understand if we are to understand anything about ourselves. We should want to know what, if anything, the sciences have to teach us. To that end, I review both the kinds of empirical studies being done and associated theoretical and polemical writings that provide a guide to their intellectual contexts.

I analyze five approaches that predominate in empirical studies of behavior: quantitative behavioral genetics, molecular behavioral genetics, social environment–oriented developmental psychology, neurophysiology and anatomy, and a set of integrative programs set up as alternatives to exclusively biological or exclusively social-environmental approaches. For contrastive purposes, I also discuss an additional interdisciplinary framework. The research tradition sometimes called human ecology has had a somewhat checkered history, likely because it had to compete with the functionalism and methodological individualism that has predominated in United States social science. And unlike genetics or neurophysiology, ecology itself has been characterized by internal debates about foundations. Nevertheless, there are a number of studies of aspects of human behavior informed in a general way by such a framework, which taken together constitute a population-level, more ethological/ecological, approach to studying behavior that is distinct from those typically engaged in the nature-nurture debate. What distinguishes this family of approaches from those I will analyze more closely is the use of properties that are only attributable to populations, such as features of social structure, to explain other population features.

The approaches with which this study is concerned are what Ernst Mayr called proximate explanations. I do not include human sociobiology or its successor, evolutionary psychology, which concern themselves with what he called ultimate explanations. In my view, the plausibility of those evolutionary accounts depends on the outcome of the research programs on more immediate causes of behavior, especially behavioral genetics, but also the neurophysiology of behavior.[23]

---

23. The success of the right kind of behavioral genetics program can only be a necessary, not a sufficient, condition for evolutionary psychology. For an interesting attempt to organize a variety of research on ag-

That is, it must be shown that the behaviors or cognitive-emotive complexes said to be adaptations can be transmitted in the way evolutionary theory requires, i.e., genetically. I am, of course, aware that there are additional research programs in psychology and in sociology directed toward understanding aggressive behavior and sexual behavior.[24] My focus will be on biological approaches focused on proximate rather than evolutionary accounts and on nonbiological approaches most closely engaged with them (either competitively or cooperatively).

Since their inception as behavioral sciences in the nineteenth century, psychology, sociology, and ethology have been characterized by multiple schools, approaches, and theories. Some are tempted, given this lack of single, governing paradigms, to dismiss them as nonscientific or, at best, prescientific. Some scholars, partly in reaction to this dismissal, have emphasized quantitative methods or formal modeling, as in much of North American sociology and economics.[25] Others have turned to biology, as a more established and methodologically sound science. But the multiplicity of approaches and attendant debates cannot so easily be avoided, since the debates also concern what is to be measured or modeled, and how and why. The label "prescientific" is, furthermore, misleading. There is a great deal of activity throughout the broad range of behavioral sciences that, by any definition, qualifies as science: various kinds of observation, measurement, and experiment. The source of discord may lie partly in the complexity of the phenomena studied, rather than in the immaturity of the research. There is nevertheless a certain value to the "prescientific," or "pre-paradigmatic," rubric, as it suggests a related way to think about the polemics in which it is invoked. The social account of knowledge and inquiry places critical discursive interaction at the center of epistemological analysis. From this perspective, one can see the polemics themselves as a means by which the scientific study of behavior is organizing itself. These debates *may* result in a unitary discipline with a single approach and single theory; they may equally result in multiple disciplines, with multiple approaches and theories. Even as the current approaches are accumulating empirical results, the debates in which their advocates engage

---

gression within an evolutionary framework, see Mealey 1995. For a thorough critique of the assumptions of evolutionary psychology, see Lloyd 2001, 2005.

24. Patrick Suppes claims that there are many more studies of aggressive behavior in the field of psychology than in biology. To some extent this may reflect issues of disciplinary location and identity. Physiological psychology, in which most psychological animal research on aggression is pursued, is arguably biological in its overall orientation.

25. Trout (1998) both describes and defends this tendency in social and behavioral sciences.

are effectively sorting out methodological, epistemological, and ontological issues.

The book is divided into two major parts. In part 1, I lay bare the structure of each of the five approaches by summarizing its theoretical/conceptual framework, giving some examples of research conducted under its aegis, and identifying questions, goals, methods, scope, limitations, and assumptions proper to each.[26] The analysis draws on research reports, research reviews and meta-analyses, and critical appraisals. A summary chapter brings together the major conclusions about each approach, dispelling the idea that there is an empirical way of settling the disputes among the proponents of the different approaches. Each generates its own set of questions, to which different forms of data, aided by different assumptions, provide distinctive answers. Even though some hypotheses supported within the different approaches seem to be in contradiction, the conflict is not, or not always, one that can be settled by appeals to evidence. Instead the disagreements are best understood as conflicts among approaches.

The second part of the book consists of three chapters devoted to philosophical reflection on the epistemological, ontological, and social issues raised by this synoptic analysis. In chapter 8, I argue that, as presently constituted, the approaches can neither be integrated nor reduced to one by elimination or reductive assimilation. Each characterizes its domain in a unique way that precludes the reciprocal evaluation suggested by their polemics. Each provides partial knowledge, conditioned on a particular characterization of what I call the causal landscape. My pluralist analysis seeks to identify what kind of knowledge each approach can offer. Unlike many other philosophers who have written about these areas of research, I am not attempting to evaluate the approaches or to argue for the superiority of one over the other. We can learn more by trying to get a handle on the complexity of the research landscape than by engaging in partisanship. Philosophy's tools may in fact be better employed in the former task than in the latter.

Analysis of the epistemological issues goes only part way toward an understanding of this research. In spite of differences in the causal or explanatory factors they investigate, these approaches share conceptions of the *explananda*, the

---

26. By "approach" I mean a complex conceptual and practical content that includes, at a minimum, characteristic questions, characteristic methods for addressing those questions, and a commitment to the importance of the questions and the answers generated by the available methods. For further discussion of this concept, see Kellert, Longino, and Waters 2006. See also Longino (2001b) on frameworks of investigation and "local epistemologies."

phenomena they are trying to explain, conceptions both of behavior in general and of the particular behaviors that are the focus of the empirical studies. In chapter 9, I analyze the ways in which the particular behaviors of interest to the studies examined are operationalized and conceptualized, and set that examination in the context of philosophical thought about the general concept of behavior. My point here is to draw attention to instabilities in these concepts, both particular and general, to the value-infused characterizations of behavior, and to the consequences of adopting different concepts of behavior as our explananda.

As already established, scientific research on behavior excites interest well beyond the laboratory. Intelligence and cognition, aggression, and sexual behavior are all of general interest to the public, to policy makers, and to workers in education, social welfare, psychotherapy, and the criminal justice system. In the final substantive chapter of the book, these consequences are made more concrete as I discuss the impact of the work on aggression and on sexual orientation on how we think about the social issues such research is called on to address. Characterizing the network of interactions among producers and consumers of knowledge poses severe challenges. How do we even identify the consumers of knowledge? Who really pays attention to the research, through what means, and how can we ascertain how it enters into their social, political, and cultural attitudes? As a first step toward gauging the presentation and uptake of this work, I engage in citation analysis of representative sets of articles, review researchers' own popular representations of their work, and consider reporting by science journalists. I suggest that the narrowed outcome of this chain of transmission interacts with needs and values in the broader sociocultural context to legitimize and reinforce attitudes that would be challenged by a more complete and nuanced representation of the research.

Conclusion

My social epistemological stance supports a pluralist response to the existence of multiple scientific approaches to a given phenomenon or range of phenomena. A nonpluralist will see multiplicity as a sign that either critical evaluation or more empirical work is needed to winnow the options to a single correct theoretical approach. A pluralist seeks instead to understand how the intellectual landscape permits the success of multiple accounts of the same phenomenon. This study can thus serve as a model for the application of philosophical pluralism in any area of science characterized by a multiplicity of accounts. Unlike most studies proposed as models in the philosophy of science, this one is focused on research whose scientific status is frequently called into question. Behavioral research has seemed lacking in comparison with research pursued under the aegis of the great unifying theories of physics. But whatever it lacks, it is surely as scientific, as engaged in observation, measurement, and hypothesis testing, as any other area of science. Only through a nonpluralist, or monist, philosophical lens can it be seen as scientifically deficient. From a pluralist perspective, it may only be a vivid example of the plurality that infects all attempts to know the natural and social world.

This study is also an inquiry into relations between science and its supporting society. The traffic goes both ways: social preoccupations influence the conduct of science and, via the way it is reported, science reinforces the assumptions underlying those preoccupations. But I will argue that the political critics of genetic research who deplore biological determinism misstate just which assumptions circulate in this way. With the benefit of the pluralist perspective, other aspects of concern come into view.

In the next several chapters, I will review the kinds of question, kinds of hypothesis, and kinds of investigative methodology that characterize the various approaches, as well as polemical interchanges among their proponents. My aim is to make clear the extent and limits of the knowledge each approach can generate. As I have said, it is not my intention to develop criticisms myself of any of these studies, but rather to understand their logical, evidential, or argumentative structure, the assumptions within which they work, the features that divide them, and the features they share. My metalevel critical focus is concerned with the representation and understanding of the relations among these approaches rather with the evaluation of any one of them.[27] My first aim is to answer these questions: What kind of knowledge of human behavior does this research offer, and what is it knowledge of? My second aim is to raise another question: Given a pluralist answer to the first, how should scientific research on human behavior inform discussions of social policy?

---

27. This disclaimer notwithstanding, I may not be able to resist pointing out difficulties in some of the claims made, especially when they concern approaches other than the one whose point of view is being represented.

Part 1

Approaches to Understanding Human Behavior

2

Quantitative Behavioral Genetics

C lassical (also known as quantitative) behavioral genetics links behavioral variation with genetic variation within a given population. One aim—for some the central aim—is to ascertain how much of that behavioral variation is due to genetic variation and how much to environmental variation. Environment is understood as encompassing socioeconomic status (SES), which includes race, education, and income levels of parents; urban, rural, or suburban location of rearing; and other such factors. It also includes professed attitudes of parents to education, discipline, and other factors relevant to childhood experience. Thus, classical behavioral genetics treats the question of interest as directly about nature versus nurture. Some behavioral genetics researchers hope to draw conclusions about individuals from such population data, while others doubt that this can be done.

Granting that both the genes and the environment of an organism causally influence its behavior, the behavioral geneticist asks what the genetic contribution to a given behavior is. The main tool used is analysis of variance, which enables identification of how much of the difference in expression of a trait in a population is correlated with genetic difference. The environmental contribution is calculated by subtracting the genetic variance from total phenotypic variance. Behavioral geneticists have introduced some refinements, discussed below, into this rough scheme. In humans, the methods traditionally used have been twin studies and adoption studies. These function as natural experiments in which the biological contribution to variation can be distinguished from the environmental contribution by treating one as constant while the other varies.

Identical twins, which result from the splitting of the zygote, share the same genome. Nonidentical twins, which result from the fertilization of two eggs in the same event of conception, do not share an entire genome and are only as genetically similar as nontwin full siblings. This fact permits several kinds of study. In studies of identical twins reared apart, the biological contribution is taken to be identical in the twins, while their environment differs. When the concordance (degree of similarity) in a given trait across twin pairs is greater than the trait's frequency in the population, the trait is deemed heritable by a figure functionally related to the measured concordance. Another kind of twin study involves comparing identical twins with fraternal twins or with nontwin siblings. When the twins or other siblings are raised in the same household, researchers can compare the concordance in expression of a given trait, T, by monozygotic (MZ) twins, who share an identical genotype, with its expression by dizygotic (DZ) twins, who share only half of their genes. The assumption of these studies is that the environment is the same for the members of the DZ twin pairs, and that only the genotype differs. A finding of higher concordance in T among MZ than among DZ twins is then treated as data supporting claims of heritability of that trait. Researchers have also compared twins reared apart with twins reared together. Here the assumption is that the environments differ for the separated twins but not for the twins reared together, while the genes in both kinds of pair remain the same. When twins reared apart show the same concordance for T as twins reared together, this too is taken to support the hypothesis that T is heritable.

In adoption studies, researchers compare adoptees with both their biological and their adoptive parents and sometimes with biological and adoptive siblings. In comparisons of concordance with biological and adoptive parents, the assumption is that the environment provided by the adoptive parents is different from that that would have been provided by the biological parents. Comparisons of biological and adoptive siblings assume that the environment of rearing is the same for both. Measured concordance in a trait across biological parents and their children in any of these comparisons, when that concordance exceeds similarities among nonbiological relatives, provides the basis for a heritability estimate. While most proponents and commentators understand quantitative behavioral genetics as applicable to differences within populations, some have argued that it can also be used to determine the genetic contribution to individual development.[1]

---

1. Burgess and Molenaar 1993.

A typical quantitative (classical) behavioral genetics study of aggression measures concordance in some measure of aggressive behavior, such as felony conviction or antisocial personality diagnosis, among biologically related family members in comparison with non- or less biologically related family members.[2] Thus researchers may compare concordances in twins raised in the same household and twins raised in separate households or in monozygotic and dizygotic twins or in adoptees and their adoptive parents versus their biological parents. For example, a twin study examining a broad range of behaviors examined concordance in measures of antisocial behavior in 331 twin pairs raised together and 71 pairs reared apart. Behaviors were identified through a self-report questionnaire (the MMPI) and included two sets of questions measuring antisocial or aggressive behavior. The concordance in answers among the twins reared apart supported a heritability estimate of .8.[3] A study contrasting behavioral concordance between adoptive parents and adoptees and biological parents and adoptees used having been listed in the Danish penal registry as the measure of aggressive behavior.[4]

Researchers Mason and Frick identified seventy such studies of aggressive behavior (that did not involve confounding issues like alcohol abuse) published between 1975 and 1991 and developed a set of selection criteria that the studies had to satisfy in order to be included in their meta-analysis.[5] They then sorted the behaviors reported in the studies into three categories: criminal offending, aggression, and (other) antisocial behavior. Each of these was then divided into categories of severe and nonsevere. According to Mason and Frick, the studies supported the conclusion that there are genetic effects on antisocial behavior and that these are stronger for severe antisocial behavior than for nonsevere. Rhee and Waldman's more recent meta-analysis is more circumspect in its conclusions, distinguishing among various operationalizations and measurement methods used in the studies included in their sample.[6] On the basis of fifty-one twin and adoption studies conducted in the framework of behavioral genetics, they propose a model that assigns exact proportions of variance to addi-

---

2. See chapter 9 for discussion of the variety of dimensions of such behavior.

3. Tellegen et al. 1988. Twin-study heritability results included in the meta-analysis performed by Mason and Frick (1994) range from 0 to .84.

4. Hutchings and Mednick 1975. This study distinguished between felony property theft and felony assault and found higher heritability for the former than for the latter.

5. Mason and Frick 1994.

6. Rhee and Waldman 2002. For a review even more explicitly supportive of behavioral genetic approaches, see DiLalla 2002.

tive (.32) and nonadditive (.09) genetic influence[7] and to shared (.16) and non-shared (.43) environmental influence. Although the largest share of variance has to be attributed to nonshared environmental influence, their comments in the discussion indicate sympathy for the view that nonshared environmental influence is largely to be attributed to genetic factors.[8] Recently Lisabeth DiLalla has proposed that one generates different heritability estimates depending on what process one uses to measure aggressive behavior: measurements based on parental ratings of their children's behavior yield lower estimates than those based on reports of nonrelated raters, whether teachers or laboratory observers.[9]

Such studies have also been performed for homosexuality. In one study, Michael Bailey and Richard Pillard recruited subjects by advertising in gay media for gay or bisexual men with either male co-twins or adoptive or otherwise genetically unrelated brothers.[10] They identified 161 individuals suitable for interview, 115 of whom had male twins and 46 of whom had adoptive brothers. The aim was to see if any difference in frequency of homosexuality correlated with a difference in genetic relatedness. They assessed the sexual orientation of the siblings both by asking the subjects what they believed about their brothers' orientations and by answers to questionnaires returned by 127 of the 135 relatives they were given permission to contact; 52% of the monozygotic twins, 22% of the dizygotic twins, and 11% of the adoptive brothers were deemed homosexual by these methods. The authors explore various subtleties in the data in their discussion. Any inferences from the data about heritability of sexual orientation depend on what base rate is assumed. The authors provide six heritability estimates using assumed rates of 4% and 10%, each paired with three possible values for the influence of self-selection in response to their advertisement. The estimates range from .31 to .74 and are in all cases higher than the proportion of variance attributable to shared environment. There are some difficulties in the data, namely that the concordances for DZ twin brothers are quite different from those for nontwin brothers. Nevertheless, the authors conclude that their study provides evidence that sexual orientation is strongly genetically influenced.[11]

---

7. Additive genetic influence sums simply with other components of variance, while nonadditive interacts with those components in nonspecified ways.

8. See the next section for further discussion of nonshared environment.

9. DiLalla 2002.

10. Bailey and Pillard 1991.

11. Bailey and Pillard use the expressions "constitutional" and "genetic" in their discussion. Interestingly, concordance for homosexuality among non-twin brothers in the sample (with sexual orientation assessed indirectly by subject report) was less than that for adoptive brothers, less than would be predicted by the

•

Proponents of behavioral genetics claim on its behalf that it can help elucidate the mechanisms underlying behavior; that it represents the appropriate extension of evolutionary, Darwinian, concepts to behavior; that it can help show the role of nongenetic, environmental factors in behavior; and that it can indicate the limits of various kinds of intervention strategies as well as the populations that may or may not benefit from such strategies.[12]

Stated broadly, the goal of research on genetic determinants of behavior is to contribute to our understanding of why individuals behave as they do. What can genetic research contribute to this understanding? First of all, the question must be revised to reflect what behavioral genetic analysis can actually accomplish. As we have seen, classical behavioral genetics draws on population genetics, which analyzes variation within populations. "Why do individuals behave as they do?"—a question about individuals—becomes "Given that population P is characterized by variation V in the expression of trait B, what explains the V of B in P?"—a question about variation among individuals in a population. Given that the focus is genetics, the question is more specifically "What does genetic variation in the population contribute to this behavioral variation?" This top down approach admittedly leaps over all the physiological and neurological activity that may also contribute to variation, but by seeking a correlate directly in the genotype, which is presumed to remain stable over time, it aspires to correlate measurable behavioral differences with biological variation that does not change over time. The presumption of a stable biological base facilitates longitudinal and multivariate analyses of twin and adoption data, as stasis and change in the heritability estimate of a trait over time or across traits at a single time are referenced to a stable genome or stable genetic variation measure.[13]

Behavioral genetics also lays part of the groundwork for extending evolutionary theory to behavior, since the finding that a given trait is heritable supports the broader evolutionary claim that the trait was selected for in some environment. But one can, especially when one pays attention to the statistical information produced, as easily see the work as showing the limits of gene action/influence,

---

genetic model based on the other data, and less than that reported in other similar studies (e.g., Whitam, Diamond, and Martin 1993; Pillard and Weinrich 1986).

12. McGue 1994; Scarr 1992, 1993; Plomin, Owen, and McGuffin 1994; Bouchard and McGue 2003.

13. The stability presupposed here is structural rather than functional, since the same structure will participate in different expression processes when activated by different enzymes or combined with other genomic sequences.

as Plomin suggests, that is, as revealing the extent of behavioral variation that is subject to influences *other* than the genetic. Given the partition of causes into genetic and environmental, simple arithmetic enables attribution of influence to environment. If the heritability statistic, *.nn*, is sufficiently low, the reasoning goes, significant environmental influence (1.00 − *.nn*) has been demonstrated.

Another value of these approaches and the meta-analyses of single variable studies in psychology is their potential to support more careful specifications of the behaviors under study. Whether distinguishing between violent and non-violent antisocial behavior or isolating components of behavior, like impulsivity, whose variation more closely tracks genetic variation, the behavior genetic approach contributes to psychology by assisting in the identification of behaviors or their components that can be studied systematically. As it comes to be used in concert with other approaches in psychology, even greater refinement in trait identification may be expected. Furthermore, it may come to be seen as contributing to an understanding of behavior without necessarily exhausting the possibilities for systematic research.[14]

Looking to a broader social context, behavior genetic research serves various kinds of social interest in particular behavioral phenomena. As stronger and stronger assumptions are used to interpret basic research findings, greater and greater claims of social relevance can be made. In spite of the role of behavioral genetics methods in determining environmental contributions to behavioral variation, research results are generally presented in terms of what they suggest about genetic influences on behavior. Both aggression/criminality and sexual behavior are, as noted earlier, aspects of social order, and there is a constant interest in controlling or regulating them. This interest contributes to determinist readings of the data in the press and by those proposing population-wide screening programs such as that suggested for fragile-X carriers.[15] Frederick Goodwin, the NIMH administrator who argued in favor of the Maryland violence conference, justified the funding of behavior genetic research into violence on the basis of the potential development of screening programs. Some commentators on

---

14. See Crick and Dodge 1996. Geneticist Atul Butte has proposed that all conditions, including psychiatric and behavioral ones, will eventually be identified and differentiated by their underlying genomic bases (see Butte 2008; Butte and Kohane 2006; and my discussion in chapter 4, note 17).

15. DeVries et al. 1998. "Fragile X" refers to X chromosomes with an unusually long CCG repeat sequence, in this case at X27.3. This repeat sequence was associated with mental retardation, hyperactivity, and some degree of increased adult aggressivity. There was debate within the treatment community as to the value of genetic screening. See Palomaki and Haddow 1993; Bonthron et al. 1993.

genetic research on sexual orientation suggest that it could form the basis for prenatal screening, while others applaud its potential to destigmatize homosexuality. On the other hand, behavioral geneticist Sandra Scarr sees the behavior genetic work as demonstrating the limited potential of social interventions. Such a stance could be seen as resistance to social engineering of any kind, especially in light of Scarr's views about basic human heterogeneity.

The multiplicity of possible pragmatic/contextual relations reflects ambivalence in the social context in which behavioral genetic research is pursued. Reading it in light of popular conceptions of genetic control, encouraged by hyperbolic claims made to garner financial support for the Human Genome Project, tends to bring out its continuity with earlier programs in eugenics, and so some proponents and most critics read it. But there are other histories (of Soviet-style social engineering, for example) in the context of which genetics research can be seen as a check against inappropriate ambitions.

Methods, Scope, and Assumptions

While the overarching question for behavioral genetics is *What role(s) do genes play in behavior B?* for quantitative genetics this resolves into questions about heritability:

- To what degree is B heritable? (How much are differences in parents correlated with differences in offspring? How much do differences in parents influence differences in offspring?)
- How much of the difference in expression of B in a population is associated with genetic difference?
- Does the heritability statistic, and hence, the degree of genetic influence on B, change over time?
- How can the methodologies used to study genetic influence on behavior be refined and extended?

To address these questions, classical human behavioral geneticists use traditional methods of ascertaining heritability: twin studies and adoption studies. More powerful experimental methods (e.g., breeding, as used in classical fruitful genetics) are not appropriate—for reasons both of ethics and of scale—for human studies. The question classical behavioral geneticists seek to answer is How much of B is genetically influenced? This "how much" does not refer directly to influences on individuals, but rather to the relation between genetic variation and trait variation in a population, which is itself measured by the relation between genetic variance and total variance. "Variance" is a technical term, referring to the average of the squared differences of each individual value from the mean population value. The genetic variance is the proportion of total variance attributable to genetic difference in a population. When carried out under the rubric of genetics, twin and adoption studies can, at best, show the extent of heritability of a trait in a given population in a given common environment. This heritability is computed from the concordance in expression of the trait across generations. Twin and adoption studies, as previously noted, permit separating concordances associated with biological relatedness (genetic similarity) from those associated with environmental similarities. With subjects categorized as genetically similar but environmentally dissimilar (twins reared apart) or genetically dissimilar but environmentally similar (adopted versus biological offspring), analysis of variance (ANOVA) techniques can be used to partition the variation in a trait between genetic and environmental factors in the context in which the trait variation has been measured. The traits examined are attributed to individuals on the basis of reports by parents and teachers, answers to personality inventories, and public records.

So the question to which the twin and adoption studies can provide answers is How much of the difference in the expression of B in environment E is heritable? The assumption is that "heritable" here is equivalent to "genetically heritable," though the methods available to classical behavioral geneticists do not permit demonstration of genetic, as opposed to other, nongenetic, mechanisms of transmission. This conclusion is possible only if one assumes that the only causes in question are genes and the environment. The methods are suitable for discriminating among and evaluating alternative hypotheses that assign different heritability quotients, that is, different allocations of the measured variance to genetic variation and to environmental variation in the population. The studies cannot, however, be generalized to answer, without qualification, the question How much of the difference in the expression of B is heritable? Both variability and heritability of a trait can change when environments change. Not only do twin and adoption studies not enable conclusions about causation in individual organisms, they support conclusions only about distributions of a behavioral trait in a given population in a given environment. Generalization beyond that population must assume similar population characteristics and similar environment. Variations on the basic pattern of analysis, however, can increase the sophistication of the hypotheses supported. Multivariate analysis, which measures the correlations among variations in different traits, can assist in the decomposition of complex phenotypic traits and also support claims of a common genetic basis for different traits. Longitudinal studies can help in identifying shifts in the relative influence of heritable and nonheritable factors over time. On the other hand, because their independent variables are conceived as genetic relatedness and nongenetic, or environmental, relatedness, twin studies are not capable of distinguishing between prenatal intrauterine effects and genetic effects, nor, because all that is known about twins is that they share a genome, are they capable of distinguishing between polygenic and monogenic traits, nor, obviously, are they able to identify genes.

Summarizing the methods employed in classical behavioral genetics, we have

- Simple heritability studies (ANOVA) that measure and compare the frequencies of some behavior B (1) in twins reared apart, (2) in MZ and DZ twins (and full siblings), (3) in adoptees and biological and adoptive parents, or (4) in adoptive and biological siblings. These measurements enable comparative estimates of relative frequencies of aggressivity or of sexual orientation in populations with different degrees of biological relatedness.
- Longitudinal studies that ascertain changes in these paired frequencies over time (e.g., whether concordance in a measure of aggression decreases or increases in more biologically related populations at different times, as compared to less biologically related populations; whether there is a direction to any change over time).
- Multivariate analysis that shows both covariance of traits (i.e., identical or similar distributions, suggesting common causality) and potential decomposability.

Criticism of classical behavioral genetics can be internal—focused on ways in which some particular study, or the interpretation of a set of studies, fails to observe the methodological prescriptions constitutive of the approach—or it can be external: directed at the entire enterprise, or at crucial assumptions of the approach. Both kinds of criticism reveal assumptions about the program—concerning, in the former case, requirements that are assumed to be satisfied by methodological precautions and, in the latter case, premises that frame and structure the program. Individual studies may be shown to fall short on methodological grounds without calling the entire enterprise into question.

One frequently voiced concern has to do with subject selection. Twin studies need to guard against the self-selection into study populations of twins who have determined themselves to be alike on some dimension of behavior. Subjects need to be recruited in a way that does not reveal the behavioral dimensions of interest and that does not in other ways result in overrepresentation of pairs of twins who are alike in these behaviors. Studies that use twin registries and other impersonal data sets, like criminal conviction, are protected from this kind of error. Studies that purport to study twins reared apart and fail to exclude twins who have been reunited after separation are not. Studies that recruit subjects from affiliative societies or from twin groups must use methods like MZ/DZ comparisons that will not bias the results. In addition, aggregation of different studies requires that the same (or significantly similar) operationalizations of a given behavior were used in measurement; otherwise it runs the risk of conflating heritabilities for different behaviors as though measuring the heritability of a single behavior.

External criticism has been directed at inferences from what well-done studies of heritability show to hypotheses of genetic influence or causation. One common misunderstanding of the behavioral genetics approach is that it purports to identify (genetic) causation of behavior in individuals rather than the basis of variation in a trait in a population. This misunderstanding is promulgated primarily in popularizations, but also sometimes in discussion by researchers themselves of the potential value of their research.[16] One common criticism, pointing out the difference between causes of traits in individuals and concomitants of variation in populations, is directed against the purported assumption that identifying correlates or even causes of variation in a particular environment is the same as identifying causes of traits tout court. This is the well-known criticism of analysis of variance articulated by Richard Lewontin.[17] There are two issues here, one proscribing inferences from a single population to the species (or whatever larger grouping of populations might be in question) and another proscribing inferences in the other direction, from a population to its individual members. In other words, it cannot be inferred from a study finding that .nn of the variance in trait T in population P is associated with genetic variance that .nn of the occurrence of T in an individual member of that population is attributable to heredity.

16. Burgess and Molenaar (1993) claim to have developed a procedure for extrapolating from population results to individuals that is essentially equivalent to the use of factor analysis in interpreting results of personality tests.

17. Lewontin 1974, repeated and cited by multiple philosophers, including R. Richardson (1984), Sober (2001), and Lloyd (1994).

Nor, as we have seen, can it be inferred from analysis only of variance for trait T in population P that the heritability statistic for T in P holds across populations. Of course, there are situations in which the latter inference can be made: those in which it has been established that the environments for all populations to which one is inferring are identical. Thus, it seems to me, the appropriate criticism is not that ANOVA can never reveal causality, but that inference from one such analysis (or a few such analyses) to a claim about causality involves empirical assumptions that must also be supported. The heritability statistic alone is not an appropriate measure of genetic contribution, because the methods for producing that statistic are not adequate to evaluate the necessary assumptions about environment. There is a further limitation of ANOVA as a method for identifying causes that Lewontin has noted. Analysis of variance requires variance. It will not identify as dependent on genes those traits that do not vary in a population, for example, being two-legged in a population of humans.[18]

The other major concern has to do with the aim of partitioning variation in behavior into one portion correlated with genetic variation and another correlated with environmental variation. Critics have pointed out that this fails to take into account gene-environment interactions: insofar as genes may express differently in different environments, heritabilities calculable in one context may not remain stable across contexts.[19] Inferences from a heritability estimate for a trait in a given environment to an estimate of heritability for that trait in general, and from that to an estimate of degree of genetic influence, are seriously misleading.[20]

The concept of environment as used in these studies poses two kinds of challenge: (1) environments may vary both over time and across space, and (2) establishing stability or variation in environments requires methods for independently identifying them. With respect to the first challenge, as just emphasized, heritability of a trait denotes parent-offspring concordance in that trait in the environment in which the trait is measured. Any extrapolation requires assuming that environments other than the one in the study do not differ from it in causally significant ways. If a study is only of a given sample population in a given environment and genetic factors interact with environmental ones, then

---

18. See also the discussion of analysis of variance in Sober 2001.

19. A stronger notion of interaction is deployed in some developmental analyses, as discussed in chapter 6.

20. The *locus classicus* for this argument is Lewontin 1974. Other versions are found in Gould 1981 and Lewontin, Rose, and Kamin 1984. The argument is repeated in Gottlieb 1995.

not only is any inference to causality or influence suspect, but so is any assertion of heritability in general. The particular degree of parent-offspring similarity in a trait ascertained in one environment may not hold in another.

The second problem threatens to undermine the very methodology of twin and adoption studies.[21] One assumption of these studies is that twins reared apart are reared in significantly *dissimilar* environments. It is simply assumed that any environment other than the one into which an individual was born will be significantly different from it. This is affirmed both in the absence of agreed-on measures of environmental similarity and dissimilarity and in the face of placement practices of finding similar homes for twins, and homes for adoptees that resemble what were or would have been their natal environments. It also ignores the possibility that the degree of environmental influence may be a function of age at adoption and ignores the similarity of uterine and early postnatal environments.[22] A correlative assumption is that adoptive and biological siblings are reared in similar environments. This treats environment as comprising gross (or shared) characteristics of the home setting, rather than individualized (or nonshared) aspects. Nonshared environmental factors like age and birth order, as well as the variation in interaction of individual personalities in different combinations, pose a challenge to this partition of the causal space.

Behavioral geneticists have developed more elaborate conceptions of "environment" to address these problems and to support their claim that behavioral genetic methods are adequate for identification of environmental influences as well as genetic ones. They first introduced a distinction between shared and nonshared environment.[23] Variation that could not be attributed either to shared genes or shared environment was attributed to the nonshared environment, which was distinct to individuals. Sandra Scarr went so far as to suggest that nonshared environment was in fact an effect of genetic variation, being social-environmental variation evoked by the individual's genetically based individuality.[24]

Other researchers approached the problem of nonshared environment by first introducing a distinction between objective and effective environment.[25] The "effective environment" is limited to variation that makes a measurable difference in accounting for variance, while the objective environment is simply

---

21. Billings, Beckwith, and Alper 1992; Lewontin 1991; Haynes 1995.

22. Properly designed longitudinal studies can address the first of these liabilities.

23. Rowe and Plomin 1984; Plomin, Owen, and McGuffin 1994.

24. Scarr 1992.

25. According to Plaisance (2006), the notion of effective environment is introduced in Goldsmith 1993.

everything that is not genetic, for example, oxygen. The concept of the objective environment corresponds to our intuitive notion of environment, but it is not the one used by quantitative behavioral geneticists. For them, what is of value is the concept of effective environment, and the distinctions that can be drawn between different kinds of effective environment. These distinctions, between what is common and what varies, are measured in terms of the kinds of patterns in phenotypic variation used to partition causes into genetic and environmental. Specifically, shared environment is defined as the (components of) effective environment that account(s) for phenotype *similarities* (in biological relatives) not accounted for genetically, and nonshared environment is defined as the (components of) effective environment that account(s) for phenotype *differences* (in biological relatives) not accounted for genetically. Thus, birth order will only be considered part of the environment if it figures in effective environment.

As Kathryn Plaisance has pointed out, the limitation of relevant environment to effective environment, along with the study-specificity of effective environment, means that no general conclusions about the relative influences of genes or environment on a trait, tout court, can be drawn.[26] The introduction of these two distinctions and their explication in terms of what is already measured by quantitative behavioral genetic methods removes the fuzziness of an unspecifiable noise or interaction term, but it does not really solve the original problem. There is still no independent way of identifying and measuring similarities and differences among environments.

Behavioral geneticists have differed in their understanding of the nature of nonshared effective environment. Scarr introduced the notion of an average expectable environment.[27] This was an environment of rearing that was not damaging. Initially she proposed treating differences in the average expectable environment as fluctuations (noise) that would cancel each other out. Later she proposed that individual variations within the average expectable environment are actually the effects of the genetic makeup of the individuals whose behavior is the object of study.[28] The idea is that genetically determined differences in individuals elicit different responses from the social, including parental, environment. Interaction between genes and environment seems to appear in the data, but really individuals create their own nonshared environments. While this

---

26. Plaisance 2006.
27. Scarr and McCartney 1983; Scarr 1987.
28. Scarr 1992.

analysis seems plausible in the case of such intrafamilial differences as whether or not a baby is colicky, the extent of its reach is not clear, especially when the variability in social and parental response to trait differences is considered.[29]

Robert Plomin, in contrast, sees the gap left between genes and the macroenvironmental factors controlled for in twin and adoption studies as causally effective nonshared environment, and proposes that one of the values of behavioral genetic methods is that they show both the limits of genetic influence and the extent of nonshared environmental influence. Using the concepts of effective shared and nonshared environment as defined above,[30] as long as the relations are additive, it is possible to parse the variation in a test population as attributable to heredity, shared environment, or nonshared environment. Plomin's proposal seems to have been adopted more widely than Scarr's by the behavioral genetics research community, although he allows that in some cases nonshared environment may be in the end attributable to genetically determined differences. What the concept of nonshared effective environment allows is a reduction in unidentified "noise" within a single study. However, it remains an additive component of variance. If the effect of genes varies with environment—that is, if there is interaction between genes and environmental factors—the concept of nonshared effective environment is not adequate to its expression.

In addition to these problems with the concept of environment, quantitative behavioral genetics must make some problematic assumptions about the relation of genes and phenotypes. Endophenotypes are intermediate physiological or psychological conditions that may be a component of or precursor to the behavioral trait of original interest. Both impulsivity and cognitive deficiencies have been proposed as endophenotypes that have aggressive behavior as one manifestation. The "one gene, one disorder" rule advocated by Plomin helps in pushing researchers to find and identify endophenotypes that have a higher heritability than the original behavior under study.[31] But it also facilitates problematic inferences even to heritability. If a trait is polygenic (i.e., involves a multitude of genes), studies on monozygotic (MZ) twins will give an overestimate of the heritability of that trait in the general population. Since MZ twins share all their genes they will share the full set of genes sufficient for expression of a trait

---

29. DiLalla (2002) embraces this notion in a review of behavioral genetic research on aggression in children.

30. Where the effective shared environment is measured by those phenotypic similarities not accounted for genetically, and effective nonshared environment is measured by those phenotypic differences not accounted for genetically. Again, see Plaisance 2006 for further discussion.

31. Plomin, Owen, and McGuffin 1994.

in a given environment. Unless the genes are linked, the probability of their co-occurrence in non-MZ siblings is much less than that of the occurrence of any single member of the set, thus yielding a greater difference in concordance rates between MZ and dizygotic (DZ) twins, with the significantly higher concordance rates for the MZ twins providing illusory support for heritability claims. "One gene, one disorder" also suggests that a trait like aggression, if determined to be polygenic, can be divided into subsets, each of which constitutes some proportion of the totality of aggressive behavior and could eventually be considered a distinct kind. The definition of behaviors is thus etiologically, rather than phenomenologically, determined.

Some developmental systems theorists have criticized quantitative behavioral geneticists for their concept of a reaction *range*. The genotype, according to some behavioral geneticists, sets upper and lower bounds on the expression of a trait, and these bounds remain stable across a range of environments; although different genotypes will express differently in different environments, they maintain the same relative ranking. The reaction-norm idea, advanced by developmental systems theorists, is instead that each genotype is "associated with a characteristic pattern of phenotypic changes in response to alterations in the environment . . . [but] the rank order of individuals [regarding degree of expression of a trait] can change appreciably and uncontrollably under novel conditions."[32] The point here is that, while across the variation in an environmental component in one setting a given genotype will characteristically be distributed into one order from lowest to highest expression, in another environment this order will not necessarily be preserved. While behavioral geneticists may not observe this refinement in practice, there is nothing in their methodology that requires that the reaction range of a genotype remain the same across different environments.[33]

Conclusion

Quantitative behavioral genetics is a study of phenotypic variation in populations, with the goal of distinguishing the (biologically) heritable portion from the nonheritable (or nonbiologically heritable) portion of that variation. The research approach involves sorting the biologically related from the non- or less biologically related in a study population and determining the frequencies

---

32. Wahlsten and Gottlieb 1997, 172.
33. See Turkheimer, Goldsmith, and Gottesman 1995.

of the trait under examination in the two groups. The ratio of the frequency in the biologically related group to the frequency in the total study population is, roughly stated, the basis for heritability estimates. Statistical analysis of variance (ANOVA) techniques are used to generate the heritability statistic. The language in which researchers express themselves indicates that they understand heritability to be an indicator of genetic causal influence on the behavior in question.

This research approach has been controversial. While it can apportion the variance in expression of a trait in a population between biological and environmental variation, a number of criticisms reveal vulnerabilities in the program. These concern the overextension of heritability conclusions and assumptions about the accessibility of relevant variation in the environment to the observation methods of the behavioral genetic approach. While its methods distinguish environments into shared or nonshared, they do not permit either more specific descriptions of types of environment or include methods of measurement sufficient to testing assumptions of cross-study environmental stability.[34]

---

34. This point has also been made by Turkheimer (2008, 2009) in recent talks.

3

Social-Environmental Approaches

Social-environmental approaches seek to understand the environmental contribution to the development and expression of various behaviors.[1] Unlike quantitative behavioral genetics, they are not intended to quantify the extent of environmental influence considered in the aggregate, but rather to identify aspects of the social environment that may have an effect on the behavior of interest. Some studies focus on the familial environment, considering, for example, parents' attitudes and interactions with their children. Other look beyond to the school environment, and peer relations, or media exposure. The methods used include correlational studies (using information gleaned from public records and from interviews and surveys), direct observation of behavior and interactions in standardized (laboratory) settings, and longitudinal studies following a set of individuals for six or more years. Typically the subjects in observational studies of aggression are young children or adolescents identified in schools or through clinics. Path analysis is used when many variables are being examined.

State, county, school, and medical records are sources of potential data for correlational studies. Cathy Widom and her colleagues employed this strategy

---

1. Anderson and Bushman (2002) identify five different theoretical orientations falling under this general designation. Because I am including these approaches as a contrast to the more purely biological approaches, I analyze them at a coarser level than do Anderson and Bushman. Hill (2002) offers a different classification in a review primarily of psychological research on conduct disorders.

in efforts to link adolescent and adult violent and antisocial behavior to abuse in childhood. Critical of earlier studies for a variety of empirical shortcomings, they sought to control for such factors as age, race, sex, and class. In one of their studies, they compared the records of 416 adults with histories of physical and sexual abuse in childhood (identified through court records from twenty years prior) with those of a matched control group of 283 adults with no documented history of abuse.[2] This latter group was found using school records. Neither the individuals thus selected who agreed to participate nor their interviewers were informed of the actual purpose of the study. Responses to the structured interviews, which included questions derived from tests for IQ, reading ability, and the National Institutes of Mental Health Diagnostic Interview Schedule, were evaluated with respect to the number of symptoms of antisocial personality manifested; 13.5% of individuals with histories of abuse qualified for a diagnosis of antisocial personality disorder, as compared with 7.1% of those in the control group. The researchers differentiated by sex, by history of criminal conviction, and by level of education and concluded from their data (and other studies in similar vein) that experience of abuse as a child is a significant causal factor in adult antisocial personality disorder. They further speculated that special prevention efforts directed toward victims of abuse could reduce later aggressive behavior.

In a structured clinical observation study, Lavigueur and colleagues sought to correlate familial interaction patterns with long-term disruptive behavior in eight- and nine-year-old boys. The boys chosen for the study were identified by teachers who completed Social Behavior Questionnaires on their students. Interactions in forty-four families were studied by observing parents with each child in question engaged in joint tasks in the researchers' laboratory. Observers used checklists in rating dyadic interactions between father and child, mother and child, and the two parents. Researchers found that negative behaviors (such as verbal abuse or attacks) and positive behaviors (such as endearments) in the parent-child dyads were not reciprocal; instead, negative behavior of one parent toward the boy was correlated with negative behavior on his part toward the other parent. Moreover, negative behavior of boys toward their mothers was correlated with fathers' negative attitudes toward their wives.[3]

---

2. Luntz and Widom 1994.

3. Lavigueur, Tremblay, and Saucier 1995. See Duman and Margolin (2007) for similar use of laboratory observation to support aggressive parents as a significant factor in offspring tendency to choose aggressive solutions.

A twenty-five-year longitudinal study conducted in upstate New York provided data to Johnson and colleagues for an investigation into factors that might mediate the association between parents' antisocial behavior and aggressive behavior on the part of their adult children.[4] The authors acknowledge multiple factors that could account for this association, including genetic transmission. Their statistical analysis of the longitudinal data controlled, on the one hand, for problematic parental behaviors by antisocial parents and, on the other, for antisocial parental behavior by problematic parents. This procedure revealed that problematic parenting (encompassing such behaviors as neglect, inconsistent enforcement of rules, and loud and/or violent arguments between parents) was a significant mediator of the association between parental antisocial behavior and aggressive behavior of adult offspring. Problematic parenting was associated with 41% of the parental antisocial behavior related to adult aggression on the part of offspring, leading the researchers to conclude that offspring aggression could have been reduced by that much had parental behavior toward their children been different.[5]

A great deal of the literature in psychology and sociology on social-environmental determinants of homosexual orientation was originally articulated during the period before the removal of homosexuality from the APA list of psychiatric disorders, when homosexuality was presumed to be the outcome of pathological developmental situations. Hypotheses such as that of the "overbearing mother" have since largely disappeared. In its place are studies relating such factors as birth order or number and sex of siblings to sexual orientation.[6] Other studies have sought to correlate experiential variables with sexual orientation. A team headed by Van Wyck and Geist interviewed 7,669 men and women (all white residents of the United States), who were assigned to one of four groups based on Kinsey scores: exclusively heterosexual, predominantly heterosexual, exclusively homosexual, predominantly homosexual.[7] These outcomes were then correlated with sets of demographic variables (rural-urban, religious affiliation, economic status, etc.), family variables (family intactness, age of parents at time of subject's birth, number and sex of siblings, etc.), gender-related variables (sex of childhood friends, participation in athletics, etc.), and variables related to pre-and postpubertal sexual experience. Multiple regression analyses were run

---

4. Johnson et al. 2004.

5. See also Brannigan et al. 2002.

6. Cantor et al. 2002. These studies, however, are often interpreted as evidence for a biological component in homosexuality.

7. Van Wyck and Geist 1984.

first, eliminating factors that neither accounted for 1% or more of the variance nor showed significant relation to outcome variables at the 0.001 level or better. This study indicated that, after gender, the factor most correlated with adult sexual preference was the character of early sexual experience. For men, learning about masturbation by being masturbated by another male or experiencing first orgasm in the course of a same-sex encounter was more highly correlated with later homosexuality than learning about masturbation in some other way or experiencing first orgasm without homosexual contact. For women, character of arousal in childhood was also correlated with adult sexual orientation but in a pattern somewhat contrary to that for men. When a girl's childhood heterosexual contact was with an adult male, the subject was more likely in adulthood to participate in same-sex relationships. Researchers concluded from this study that character of childhood sexual experience was the most influential factor, after gender, influencing adult sexual orientation. Other researchers have emphasized the plasticity of sexual orientation among women, suggesting a greater sensitivity to both history and immediate social context.[8]

Typically, in such studies, children are identified as aggressive through some combination of teacher ratings, peer ratings, parent ratings, and self-report. These are obtained using standardized inventories, checklists, questionnaires, or interview schedules. Some of these simply differentiate between aggressive and nonaggressive youths, others differentiate degrees of aggressivity, and others employ psychiatric diagnostic categories (such as childhood conduct disorder and oppositional defiant disorder) from the American Psychiatric Association's *Diagnostic and Statistical Manual of Psychiatric Disorders*. Researchers then seek to correlate the behavior either with parental attitudes and practices or peer relations, as identified by the same means, or with directly observed interactions. In the latter case, as in the Lavigueur group's study, children and parents are brought into a clinical or academic psychological setting and asked to perform a series of activities. The interactions, between parents and between parents and children, are observed and recorded using a standardized protocol. Such studies have attempted to establish the relation of parental disciplinary practices and the level of harmony or discord within a family (ascertained via self-report) or the relation of observed positive and negative interactional behaviors (for example, endearments or verbal attacks) with behavioral problems.[9] In a 1995 review article, John Constantino examined epidemiological, behavioral-genetic, and child

---

8. Peplau and Garnets 2000; Diamond 2008.
9. Speltz et al. 1995.

development studies to identify attachment deficits in early childhood as a key factor in later aggression.[10]

As if in response to the call for more finely grained studies, Beauchaine and colleagues investigated specifically the responses of mothers to their children's misbehavior. Aggressive boys were recruited from a local clinic and nonaggressive boys from a local Head Start program; mothers of the boys identified as aggressive were found to be more coercive and to offer fewer explanations of their responses to misbehavior and fewer alternatives to their sons' perceived misbehavior than mothers of nonaggressive boys.[11] A study of mother-infant interactions correlated the aggressive behavior of two-year-olds with forms of maternal response; both overreaction and neglect were correlated with higher levels of aggression on the part of the toddlers.[12] Other researchers have sought to coach parents in alternative interactional styles or disciplinary practices and to measure the effect on child and adolescent behavior of such changes in parental behavior.[13] And some have investigated the relation between aggression and peer relations.[14] A recent study investigated the effectiveness in reducing delinquency of a range of psychosocial interventions for use in the juvenile justice system.[15]

It is obviously much more difficult to establish a relation between early childhood experience and late adolescent and adult behavioral patterns, since it is difficult, except in a genuinely long-term longitudinal study, to establish with any reliability what childhood experiences in the distant past were. Several studies have got around this obstacle by using court records to identify victims of childhood abuse and neglect and either interviews or subsequent court records to ascertain the incidence of antisocial personality disorder or adult criminal offending in those populations.[16] The New York state study by Johnson's group used data drawn from a twenty-five-year longitudinal study of multiple aspects of development.

Public databases are less informative as sources for studies of sexual orientation, which is attributed almost exclusively on the basis of self-report, while environmental factors that might have played a role are identified using biographical questionnaires administered to subjects. Questions in surveys or longitudinal

---

10. Constantino 1995.
11. Beauchaine et al. 2002.
12. Del Vecchio and O'Leary 2006.
13. McCord et al. 1994.
14. Bierman, Smoot, and Aumiller 1993; Bierman and Smoot 1991; Miller-Johnson et al. 2002.
15. Sukhodolsky and Ruchkin 2006.
16. Luntz and Widom 1994; Rivera and Widom 1990; Widom 1989b, 1989c.

studies designed to generate information about sexual orientation may fail to do so because of internalization of homophobia or caution induced by the continuing stigmatization of gays.[17] As is the case for some genetic studies, potential subjects are recruited through advertisements in publications oriented to a gay readership, on notice boards in gay social venues (bars, clubs, or campus organizations). Subject populations are limited, then, to individuals who are willing to self-identify as gay. One hypothesis investigated recently concerns the relation between a bad or abusive relationship with one's father and later homosexuality in men. A 1989 study by Joseph Harry, with subjects recruited through fliers distributed at social events sponsored by a campus organization for gay students, indicated higher rates of childhood abuse at the hands of fathers among the gay students than among the heterosexual controls.[18] As acceptance of homosexuality has increased in the academy and in clinical professions, the focus has shifted from inquiring into factors that play a role in determining sexual orientation tout court to inquiry into variations in behavior or experience within gay populations. Thus, more typical of recent social-environmental investigations are studies that attempt to correlate some feature of childhood or ongoing social experience (e.g., health, attachment behavior, or suicidal behavior) with variations in the behavior of gay youth.[19]

•

We might characterize the overall purpose of environmentally oriented behavior research as understanding the variety of and interaction among social-environmental factors that affect human behavior. Advocates of environmental approaches are much more willing to claim social relevance and a social purpose for their work than are biological researchers. Based on their studies of the relative influence of various factors, researchers develop programs for intervention in schools or to teach parents alternative communication styles.[20] For example, the researchers in the Lavigueur study speculated that coaching parents in alternative styles of interaction could reduce the chances that their child's disruptive behavior would later develop into more serious antisocial behavior.

Some researchers clearly assume that parents' behavior has a causal effect on

---

17. For discussion, see Fay et al. 1989; Anderson and Stall 2002; Villaroel et al. 2006.

18. Harry 1989.

19. Bontempo and d'Augelli 2002; Plöderl and Fartacek 2009; Landolt et al. 2004.

20. See, e.g., Lerner 1998; Haapasalo and Tremblay 1994.

that of children.[21] Others acknowledge that in the networks of association their studies uncover, causal relations might not have the effect they assume.[22] Widom's early review of studies of the social transmission of abuse is an example of a disconfirming study pursued within the social-environmental approach.[23] Widom reviewed case records to show that the widely accepted notion that childhood abuse results in higher rates of delinquent and criminal behavior was not borne out by the data that had been collected to date. She criticized the work for depending too much on self-report and retrospective data, for failing to distinguish between neglect and abuse on the part of parents, and for failing to distinguish among problematic behaviors of the subjects. She went on to design studies that met higher methodological standards, which, once she (and others) had carried them out, allowed her to demonstrate a connection between abuse and later delinquency.[24]

Diana Baumrind takes a somewhat equivocal (perhaps strategic) position with regard to causality.[25] She accepts that causal inferences are not licensed by the correlational data produced in social-environmental studies. Such data are, however, valuable for eliminating, even if not confirming, hypotheses. Furthermore, she claims, it is reasonable to conclude on the basis of a number of studies that parental behavior does influence children's development. This being the case, these environmentally oriented studies are valuable because they can (help to) identify children "at risk" for antisocial behavior or delinquency and to identify points at which to intervene in family and school dynamics. In direct response to behavioral geneticist Sandra Scarr, who, as we have seen, treats environment as a given, Baumrind urges the use of environmentally oriented research to alter the "normal" range of social environments, that is, to improve school and home environments at the lower end of resource distribution scales. And, in a meliorist declaration of faith, she says that even if heritability accounts in large part for what is, it doesn't follow that social interventions can't improve on that. The implication is that social-environmental research approaches are required to identify which social interventions will be effective.

A further purpose has less to do with establishing causality than with identifying aspects of the experience of behavioral groups that may warrant special attention. Thus, the Harry study, if significantly replicated, would warrant thera-

---

21. Haapasalo and Tremblay 1994.
22. Bierman and Smoot 1991.
23. Widom 1989a, 1991; Rivera and Widom 1990.
24. Luntz and Widom 1994.
25. Baumrind 1993.

pists and counselors inquiring into parental relationships, especially possibly abusive relationships, when treating clients who happen to be gay.

Considered in a broader social context, this research, whatever specific hypotheses it confirms or disconfirms, tends to enhance the authority of social work and clinical psychology, as distinct from the medical professions, with respect to understanding and management of the behaviors investigated. By refining the environmental and familial variables examined, it guides social workers and clinical psychologists in identifying youth who are thought to be at risk. Studies of specific interventions inform these professionals as to what kinds of strategies are effective and what kinds are not. And noncausal, correlational studies may reveal the concentration of certain kinds of experience among certain groups and connect these to heightened risk for other conditions. Thus, researchers tend to keep one eye on transformations of treatment or policy in schools, clinics, or juvenile justice venues. While they have little input into the design or goals of those policies, at least part of the policies' legitimacy and social workers' authority in their implementation lies in their being informed by behavioral research.[26]

Methods, Scope, and Assumptions

At their broadest, the social-environmental approaches ask the question *What role do environmental and other exogenous factors play in behavior B?* This question seems parallel to the question asked by behavioral geneticists, but while quantitative behavioral genetics purports to assign proportions of the variance in a population to genes and environment, respectively, environmentally oriented researchers ask questions about the character of those environmental influences:

- What role do gross or macro-level social variables (social class; ethnic, racial, and cultural identity; urban, suburban, or rural; immigrant or native status) play in the expression/frequency of B? Does one or more of these predominate in the expression of B?
- What role do micro-level variables (family, school, peers, media exposure) play in the expression of B? Does one or more of these predominate in the expression of B?

---

26. This is increasingly the case given emphases on "evidence-based policy." See Cartwright (2006) for discussion of the pitfalls of this demand.

- Does the influence of micro-level variables in the expression of B vary in relation to macro-level variables? How do they interact?
- How do differences within a family influence the expression of B by its members?
- How can familial interactions relevant to B be studied?

In answering these questions, the social-environmental approaches use measures of aggression similar to those used by behavioral geneticists, except that some studies treat a continuum of social behaviors, from the antisocial to the prosocial, rather than limiting themselves to the antisocial. Prosocial (or "sociability") behaviors include offers of assistance, participating in cordial verbal exchange, and display or exchange of affectionate gestures. Aggression is measured by behaviors both physical (hitting and starting fights) and verbal (angry, hostile speech), ascertained by self- and other reports; delinquency, ascertained in court records; psychological classifications like antisocial personality disorder and childhood conduct disorder, ascertained through psychiatric diagnosis; or hostile, confrontational interactions, ascertained through direct observation.

The social-environmental approaches in psychology seek to associate distributions of one or another of these measures of aggression with variation in some environmental factor, whether parental behavior, educational experience, media exposure, or peer relations. Most of the studies using questionnaires or interviews report using more than one measurement strategy (e.g., both self- and other report, or both self-report and court records) to enhance the reliability of behavioral ascriptions. The methods employed are both retrospective and prospective. Retrospective methods are employed with populations whose relevant parameters are determined via interview and questionnaire or via direct observation. These include

- comparing the distribution of a postulated causal factor (abuse and neglect, foster home placement, socioeconomic status, disciplinary practices in the home) in the target population with its distribution in a control population;
- comparing the distributions of a variety of possible causal factors in target populations with those in control populations; and
- path analysis to ascertain the strength of links among multiple factors.

Prospective methods involve introducing some change into a sample of the target population and determining its effect. These include

- teaching alternate forms of dispute resolution and ascertaining short and long-term changes on some measure of aggression, and
- coaching parents in alternate forms of discipline and observing changes in parent-child interactions or in some aggression measure in affected children.

Controls can be either individuals in a given population matched with study subjects in all but the trait being studied or individuals who have not experienced the environmental variable in question. In the former case, the hope is to find a common (or common enough) environmental factor in the background of the study subjects that is significantly less common in the background of the controls. In the latter, the hope is to find some common effect of the environmental variable being studied (often a reduction in problematic behavior subsequent to interactional retraining).

Environmentally oriented researchers have been accused of valuing political correctness over knowledge and of permitting their concerns for social justice to interfere with their science.[27] This leads them, it is said, to select possible causes for investigation based on their perceived manipulability and to ignore genetic issues out of fear of their social meaning. They have also been criticized for assuming that there is a direct connection between science and politics.[28] More scientifically relevant is the concern that they fail to control for the possibility of a common cause for the early experience and the later behavior. Furthermore, they do not include genetically relevant information in their data, as they do not distinguish between biological and adoptive families.[29] This, it is said, leads them to confound genetic transmission with socialization. Where biological causes are separable from social ones, the behavioral geneticist claims against the environmentalist that biology is a better predictor of similarity and dissimilarity in behavior than social factors.

A number of authors have reanalyzed social-environmental studies to demonstrate this point. McGue rebuts the hypothesis that parental divorce is a major environmental factor in the divorce rate in offspring by comparing divorce rates for monozygotic twins in the Minnesota Twin Registry with those for

---

27. Scarr 1994.

28. Weinrich (1995) argues that the political concerns of opponents to biological research on sexual orientation are misplaced and that these critics misunderstand the biological approach.

29. DiLalla and Gottesman 1991.

dizygotic twins.[30] This comparison, he claims, suggests that the similarity between parents and offspring is a function of genotype, not of parents' failure to
model stable marital relationships. Scarr reanalyzes several studies of relation
between parental rearing styles and children's school achievement to show that
the factor most highly correlated with children's performance is parental IQ and
education.[31] What mediates parents' rearing styles with their children's school
achievement, then, is the heritable element of IQ. The clear implication of both
of these writers is that other socialization studies will show the same defect: restriction of study variables to environmental factors fails to reveal the most likely
causal factor—genes. DiLalla and Gottesman make a similar point in disputing
Widom's hypothesis that experiencing abuse as a child is a good predictor of
later delinquency or criminality.[32] They take Widom to task for omitting consideration of the behavior genetic literature in her review of the socialization literature. In their view, the twin and adoption studies showing a high heritability for
criminal behavior provide a framework for understanding whatever parent-child
similarities in abusive behavior have been found, and these should have been invoked as the obvious explanation. Widom's response to this criticism is to point
out, first, that her focus was on the consequences of child abuse, not the causes
of antisocial behavior and, second, that, while multiple etiological factors are
indeed likely implicated, DiLalla and Gottesman have exaggerated the role of
physiological and genetic factors relative to the evidence available.[33]

Where behavioral genetic researchers tend to see their work as refuting or
blocking claims regarding the role of gross socioeconomic status (SES) factors
or of factors common to all members of a household, many socially oriented
studies investigate more finely grained features of the environment (i.e., what
is called the nonshared environment). Their research design is intended to discriminate among potential causal factors within the environment, not between
genetic and environmental contributions to behavior. Researchers studying
and intervening in social and familial interactions may, but need not, assume
that these factors are causally independent from properties or behaviors of the
subjects whose environment they constitute. Since many interventions focus
on changing parental behavior, what behavioral genetic researchers would re-

---

30. McGue 1994.
31. Scarr 1997.
32. DiLalla and Gottesman 1991.
33. Widom 1991.

gard as nonshared environment is, however, effectively treated as independent. Baumrind defends this focus on the grounds of the adults' greater power in parent-child relationships.[34] This, of course, assumes that the adults are not unconsciously responding to features of the children's behavior or personalities, an assumption directly contradicting Scarr's proposal that nonshared environment (i.e., those parental behaviors) can be treated as a genetic effect.[35]

Just as genetically oriented researchers must assume uniformities of effective environment (e.g., the "average expectable environment"), environmentally oriented researchers must discount genetic variation in their subjects, assuming that subjects are sufficiently endogenously uniform or that their genetic variation averages out and does not interact systematically with the environmental/experiential variables being studied.[36] If the aim of social-environmental approaches were to distinguish between environmental and genetic causes of behavior or to partition behavioral variance into that attributable to shared genotype and that attributable to environment, this feature of the research would indeed be a weakness. But these researchers' methods are directed toward distinguishing among possible environmental factors, not between genetic and environmental factors. Large samples sorted for sex and SES are employed, with matched control populations, to eliminate some confounding variables, but these tend to be the same sorts of causal influence (e.g., family SES) whose role the genetically oriented researchers are also seeking to rebut.

What these studies can show is the relative strength of a particular social or environmental factor in comparison with others in the same general setting. They tend to assume either uniformity or random variation of genetic or other individual, internal factors (such as patterns of hormonal or neurotransmitter secretion) and thus purport to measure the relative strength of the various kinds of social factor—parental interaction, parental discipline, parental attitude and behavior toward other family members, peer support, school environment—in producing a particular behavior.[37] Typical hypotheses studied, then, would be, in retrospective studies, "Parents who were themselves abused as children are more likely to engage in emotional or physical abuse of their children" or, in prospective studies, "Coaching parents to adopt communicative style C in situ-

---

34. Baumrind 1991, 1993.

35. DiLalla (2002) also endorses this understanding of nonshared environment.

36. Sesardic (1993) highlights such assumptions in an attempt to overturn widely shared reservations about behavioral genetics.

37. When Harris (1999) controversially disputed the efficacy of parenting on chidren's outcomes, she contrasted parenting with peer influence, not with genetic influences, for example.

ations S reduces the frequency of aggressive responses from children." The hypotheses that can be empirically compared by these methods concern alternative potential environmental factors. Since the influence of any such factor may vary, given variation in other environmental factors, and since the magnitude of their population-level effects may vary as and if genomic distributions vary from population to population, the conclusions of social-environmental studies are as limited in generalizability as the conclusions of classical and molecular genetics.

Conclusion

Social-environmental approaches to understanding human behavior, pursued in sociology and psychology departments as well as in clinical settings, seek to understand the contribution of social experience to the formation of behavioral dispositions or to identify correlates of behavioral dispositions and their expression that are relevant to therapeutic contexts. Methods range from the analysis of large databases to survey and interview to laboratory observation and experiment. Aggressive behavior, in one or another operationalization, can be linked to maltreatment as a child or to forms of parental communication. While some research on sexual orientation has shown a relation between the character of an individual's first sexual experience and later erotic orientation, more recent research tends to focus less on etiological factors and more on the contribution of early socialization to behavioral/experiential variation among gay individuals.

Like the other approaches, the social-environmental approaches must assume that they have identified a coherent phenomenon for study, that is, that the definition and operationalizations of aggressive behavior and of sexual orientation pick out a stable class of phenomena into whose etiology or about whose correlates it makes sense to inquire. As in quantitative behavioral genetics, researchers must make assumptions about the causal landscape in which they are working, in particular they must assume that potential causal factors other than those they are measuring are either common to the individuals in the subject populations under study or that they are randomly enough distributed not to bias the results. When using correlations to support causal inferences, when the information is from a Humean or Millian perspective incomplete, they must also assume (as Baumrind acknowledges) that the correlated factor has causal efficacy. The point of a study is to connect specific causally efficacious factors (or factors believed to be causally efficacious) with specific outcomes.

4

Molecular Behavioral Genetics

ust as quantitative behavioral genetics seeks to adapt the methods of classical genetics to the study of human behavior, so molecular behavioral genetics seeks to adapt the methods of molecular genetics. Since the 1953 identification of the structure of the DNA molecule, techniques enabling identification, isolation, and manipulation of genetic material have developed at an amazing pace, spurred on by the commercial and medical potential of genetic and genomic technology. Whereas quantitative behavioral genetics was limited to inferences based on observation of transmission patterns of phenotypes, the ambition of molecular behavioral genetics is to link specific DNA sequences or chromosome loci to behaviors—to find, that is, direct associations between genetic configurations and traits.[1] Molecular behavioral genetics has aimed to identify, at whatever grain of identification is possible at the time, gene regions that might reveal associations of specific alleles with behavior. Traits, whether physiological or behavioral, common to males in a maternal line, have been especially amenable to study. Males have only one X chromosome, while females have two. A trait expressed in males can be assigned to the X chromosome by review of a

---

1. The effort to attribute (at least some) aggressiveness as a function of possession of an extra Y chromosome might be considered an intermediate between classical and molecular genetic approaches to behavior. In spite of continuing interest in identifying a role for the idiochromosomes in sex-linked behavior, the XYY research has been pretty thoroughly discredited. See Wasserman and Wachbroit 2001; S. Richardson 2009.

family pedigree. If males linked through a maternal line share a trait, X-linkage is suggested.[2] This narrows the region to be searched to one chromosome. Already established associations of physiological conditions with behaviors are also a clue; the association of variations in serotonin metabolism with aggression has been one source of clues to genetic precursors of aggressive behavior.

In one of the early investigations using molecular genetics methods, researchers zeroed in on a Dutch family, fourteen of whose male members had engaged in repeated episodes of unprovoked aggressive behavior. Eight of the fourteen, including the five exhibiting the most extreme levels of violence, consented to be studied by H. G. Brunner and his associates. Blood samples were obtained from twenty-four family members. The eight exhibiting violent behavior were found to share allelic variation on a region of the X chromosome coding for MAOA, the A form of the enzyme monoamine oxidase, which is involved in the metabolic cycle of serotonin.[3] Reduced urinary excretions of metabolites of MAOA activity further implicated the MAOA locus, as distinct from other polymorphic loci on the X chromosome. Brunner and colleagues concluded that the structural abnormalities in the gene for MAOA constituted the genetic input into the abnormal behavioral pattern in this family. The Brunner study stimulated much concern over possible genetic intervention and genetic discrimination.[4] This has subsided, and studies of the roles of irregularities in genes related to MAOA and to other components of serotonin metabolism have proceeded apace. For example, in 1999 Manuck et al. began publishing results of their research on the role in aggressive behavior of a polymorphism in the gene for another component of serotonin metabolism (tryptophan hydroxylase). In 2000 they published work on polymorphisms in the MAOA gene that associated allelic variation in a promoter region of the gene with variation in aggressive behavior. In both studies,

---

2. Hemophilia, an example of such an X-linked trait, is concentrated among male relatives on the maternal line.

3. Brunner, Nelen, Breakefield, et al. 1993; Brunner, Nelen, van Zandvoort, et al. 1993. Five members of the family exhibited extreme levels of violence, while nine others exhibited more moderate, but still higher than average levels of violence.

4. Simm 1994. Ironically, Brunner (1995) issued a caution regarding overinterpretation of the study. The MAOA gene (or rather its variant) can't be called an "aggression gene," both because of the restriction of the study to one family and because of the complexity of behavior compared with the simplicity of a gene. While molecular biology has since demonstrated the complexity of gene activation and of the genome and genetic processes generally, no one then holding Brunner's view would be compelled to change it now. In a development he would surely deplore, the MAOA allele has now been dubbed "the warrior gene" by physical anthropologists studying the association between MAOA and aggressive behavior in nonhuman primates (Gibbons 2004).

interview and questionnaire data were used to distinguish aggressive from non-aggressive individuals.[5]

And in 1993 Dean Hamer and colleagues at the National Institutes of Health published the results of a study seeking to identify genetic loci associated with homosexuality. They invoked twin studies, as described in chapter 2, as well as other population genetic work to support the initiating hunch that there is a genetic component to sexual orientation. They recruited a study sample of 76 men from HIV clinics and gay organizations in the Washington, DC, area; adding family members brought the number of individuals in the study to 122. Analysis of a separately recruited group of thirty-eight brother pairs showed an incidence of homosexuality in brothers of gay men that was 6.7 times higher than would be expected from the estimated base rate of 2%. Family histories on members of the main sample also showed a higher concordance rate among gay men and their maternal uncles and male cousins who were the sons of maternal aunts (7.5%).[6] Again, this incidence is significant if one assumes a background rate in the general population of 2%. A more refined sample revealed higher concordances (10.3% and 12.9%, respectively). Generally, the pedigree analysis showed higher concordances in the maternal descent lines than in male descent lines, and this disparity suggested X chromosome involvement. DNA was obtained from forty sibling pairs. Using polymerase chain reaction (PCR) techniques, the researchers surveyed twenty known polymorphic sites on the X chromosome for any commonalities among the sibling pairs. This revealed common allelic markers in each pair at the Q28 locus of the X chromosome.[7] This finding led the researchers to conclude that genes play a significant causal role in the development of homosexual orientation in (some) gay men. Efforts to replicate the Hamer findings in similar but different populations have failed.[8] While some see this failure as refuting claims to a genetic basis for homosexual orientation, others see it as a reason to pursue the issue with improved or different molecular technologies.

---

5. Manuck, Flory, Ferrell, Dent, et al. 1999; Manuck, Flory, Ferrell, Mann, and Muldoon 2000. See also Guo et al. 2008; Bernet et al. 2007. Work on MAOA is the basis of much of the work by Caspi and Moffitt to be discussed in chapter 6.

6. Hamer et al. 1993.

7. Thirty-three pairs were concordant for all five markers in the XQ28 region, while 7 were concordant in at least one, but not all five. As each locus had multiple variants, what each pair shared was the property of having an identical allele at that locus. It is not claimed that there was one variant shared by all the subjects whose DNA was tested.

8. Rice et al. 1999.

Molecular behavioral genetics has employed a variety of investigative technologies. The most common is linkage analysis, as used in the Hamer study. Developed for use in medical genetics to investigate such conditions as cystic fibrosis, linkage analysis looks for common genetic markers in biological relatives identified as sharing the same trait. Some chromosomal loci are distinguished by having multiple possible alleles that vary by one or more nucleotides or that have multiple distinctive repeat sequences. Systematic variations at multiallelic loci can be more easily identified in cloned DNA samples and are known as markers. Linkage analysis in behavioral genetics associates allelic variation with behavioral variation. While the shared variation at a locus, the marker, is not itself the genetic variant causally involved in expression of the trait, its being shared by individuals who also share the trait is taken as evidence that a gene in that region of the chromosome is involved with the trait. When a physiological intermediary can be associated with a trait, and the genetic precursor is known, it is possible to establish a much closer relation between a behavior and a genetic configuration. This was the case with the Brunner study, which proceeded via already established associations among MAOA levels, serotonin metabolism, and aggression to identify the higher number of repeats in the promoter region of the MAOA gene with the behavior of the members of the Dutch family under study. As techniques for sequencing genes and analyzing the chemical composition of the genome have grown in power, more studies have been performed establishing connections from traits, including behavioral ones, to more finely grained characterizations of a putative genetic component. For example, polymorphisms in a single nucleotide (SNPs) can be identified and associated with phenotypic traits.

Both the Brunner and Hamer studies, conducted in the early 1990s, required some way to limit the region of the genome surveyed. Now that the entire human genome has been sequenced, it is possible to scan the entire genome for significant associations to phenotypic traits. Genome-wide association studies (GWAS), in conjunction with fine-grained chemical analysis, can identify sets of genes (or gene regions) associated with a given trait. Given the current consensus that most behavioral phenotypes that are genetically influenced are polygenic, GWAS has seemed promising as a strategy for identifying collections of genes. One such study revealed several chromosomal regions associable with male homosexuality.[9] The associations were suggestive, but not definitive

---

9. Mustanski, DePree, et al. 2005.

because of deficiencies in their statistical power.[10] Recently a team, including Hamer, has investigated a possible relation between patterns of X-chromosome inactivation in mothers and the sexual orientation of their offspring.[11] Here the idea is that each female receives one X chromosome from her father and one from her mother. One or the other of these tends to be inactivated in each cell. Variation in the distribution of paternal Xs and maternal Xs in a woman is hypothesized to have an effect on her offspring, in this case, on their sexual orientation. While neither study is definitive, they indicate paths forward, and show how researchers on the genetics of human behavior continue to employ the latest experimental techniques. As understanding of the complexity of the genome continues to develop, additional investigative strategies are being pursued. Several epigenetic mechanisms that either enhance or inhibit gene activity have been implicated in behavioral changes in rats.[12]

Human molecular behavioral genetics, like human quantitative genetics, is hampered by the fact that the kinds of manipulations that would be more probative of genetic causation would involve intervention in reproduction that is morally off-limits. Nevertheless, studies on other animals, from drosophila to nonhuman primates, buttress the human studies and suggest associations that might be pursued using methods available to human geneticists. In a research process known as reverse genetics, researchers introduce mutations into the genome and identify the consequent variations in behavioral routines. These studies can be used to suggest mechanisms of gene expression as well as behaviors in other organisms (e.g., humans) that may be subject to genetic influence. Several studies using heat stress to induce mutations in gene regions of drosophila have reported that the resulting male flies tend to form rings of multiple individuals connected one to another in the position characteristic of copulation. The researchers speculated that this research might provide a clue to human homo-

---

10. For a description of a successful GWAS study, see Shuldiner and Pollin 2010. This was effectively a meta-analysis that pooled a number of GWAS studies, ultimately resulting in a subject pool of 100,000 individuals; even then, only about 10% to 12% of the presumed genetic variation was accounted for by genes identifiable through the GWAS. Given the much smaller sample sizes used in behavioral studies, it is hard to see that GWAS will yield more robust results than linkage analysis. McClennan and King (2010) offer an equally pessimistic assessment of GWAS possibilities for mental illness, as does Turkeheimer (2012) for any complex traits. And Austin et al. 2006 suggest that even when associations are found they should be viewed with extreme caution.

11. Bocklandt, Horvath, et al. 2006.

12. Miller 2010. This epigenetic work is better understood as exemplifying integrative research approaches. For further discussion, see chapter 5.

sexuality.[13] Epigenetic research on rodents has likewise been presented as potentially extendable to human behavior.

•

Molecular behavioral genetics has both bottom-up and top-down goals. The top-down goal starts with a particular behavioral pattern and seeks to learn its genetic antecedents. This parallels the strategy of medical genetics, which seeks to identify the genetic antecedents of specific diseases, such as cystic fibrosis and particular cancers. As is the case for those diseases, only some of the individuals exhibiting a particular behavior turn up as carriers of any given candidate gene (such as the MAOA gene). Nevertheless, finding a genetic antecedent for even a portion of the incidence is a start toward identifying at least one pathway of the condition's development.[14] The bottom-up goal involves the contribution behavioral genetics can make to the general question about the role of genetic material and genomic processes in shaping organisms' overall functioning, of which behavior is one aspect. This research can extend understanding of genes, gene action, and gene activation. Of course, associating genetic mutation and behavior links two endpoints in a long causal chain and constitutes only a first step toward understanding the full range of developmental processes and of organismic function.

Top-down and bottom-up goals involve different research methods that produce different kinds of knowledge. Research driven by top-down goals starts with a distribution of a behavioral trait in a population and associates a genetic configuration with some portion of that distribution. Bottom-up research starts with a distribution of a genetic configuration in a population and associates a behavioral outcome with a portion of that distribution. Here reverse genetics, such as Jeffrey Hall's work on the behavioral genetics of fruit flies, provides a model.[15] This work involves introducing specific genetic mutations and observing the consequent effects in behaviors or motor patterns, rather than starting, as is conventional, with variations in trait expression. Given the obvious constraints against intervening in the genome of humans, this bottom-up goal is achieved via

---

13. Zhang and Odenwald 1995. As an example of use of such research as an entry point to physiological processes, see Kitamoto 2002.

14. As Schaffner (1998) has emphasized, only when a complete biochemical and physiological pathway from genetic configuration to phenotypic condition is identified can it be said that a genetic explanation has been produced.

15. Hall 1994.

research on animal models. Occasionally, an "experiment of nature" will appear, such as a behavioral cluster associated with individuals known to be related and sharing a gene whose immediate role in the organism is at least partially known. Such cases, like the family in Brunner's study, are few and far between, but they provide hope of an integration of top-down and bottom-up analyses.

Molecular genetics is prompting some conceptual refinement through the "one gene, one disorder" approach, which breaks down a complex (or composite) trait like mental retardation into distinct (but similar) traits individuated by distinct etiologies.[16] "One gene, one disorder" has its complement in the multivariate analyses used in quantitative behavioral genetics to identify co-occurring traits that may be constituents, in whole or in part, of some more grossly identified trait. These approaches lie behind the identification of impulsivity as a lower-level disposition that may be a heritable and genetically based component of aggressivity. Thus, the assumption that there is a genetic influence on behavior drives both traditional and molecular researchers to refine their concepts to develop "constructs" that can be more reliably associated with genes. That is, the general assumption drives a search for ways to realize the subordinate assumption regarding the individuation and identification of traits.[17]

Proponents of molecular behavioral genetics claim that it can help elucidate the mechanisms underlying behavior; that it can indicate the potential of biochemical and other intervention strategies and identify populations that may benefit from such strategies; and that it provides the basis for a coherent classification of human diseases. From the perspective of a strong genetic ideology, however, this research also extends a model of genetic control of organismic function from physiology to behavior. This ideology is less characteristic of the researchers themselves than of propagandists for the Human Genome Project, of science journalists who need a simple model to provide a context for the research results they describe, and of health and social policy analysts who see

---

16. Plomin, Owen, and McGuffin (1994) use the expression "complex" for traits that turn out to be families of symptomatically similar but distinct repertoires with distinct etiologies, such as the multiple forms of mental retardation.

17. Atul Butte has, as noted in chapter 2, claimed that genetic etiology ought to be the basis for classification of physical *and* psychological disorders (Butte 2008; Butte and Kohane 2006). The hope would be for better treatment options. Such a proposal seems to underestimate both the extent to which relevant gene sequences are pressed into service by nongenomic factors and the extent to which many conditions are so polygenic as to make such identification as complicated as current classifications are vague. In addition, if taken up, such a nosological practice might delegitimate diagnoses that cannot be traced to genetic configurations. Liu, Wise, and Butte (2009) advocate a gene-environment classification system that seems to take account of some of these concerns.

problems like social violence and criminality as medical issues for the perpetrators at least as much as for their victims.

Methods, Scope, and Assumptions

The questions molecular genetics purports to answer, then, are these:

- Can any genetic markers be associated with the incidence of some behavior B in a given pedigree?
- Can a specific allelic variation or set of allelic variations be associated
  with the incidence of B?
- Can such an allele or set of alleles (or neighboring alleles) be shown to
  be causally implicated in the incidence of B?

Various methods are employed to answer these questions:

- Pedigree analysis to find biological relatives with the trait under study
  (aggressivity, sexual orientation), to identify patterns of relatedness, and
  perhaps to identify other shared phenotypic traits (e.g., in serotonin
  metabolism) to help in the search for markers.
- Gel electrophoresis to identify gene sequences; PCR to rapidly replicate
  a DNA sequence of interest; and SNP to identify fine-grained chemical
  variants at the nucleotide level.
- Linkage analysis to identify genetic markers whose allelic variation
  matches the behavioral variation in the pedigree.
- Genome-wide scanning to identify genomic variation associable with a
  behavioral variation.
- Reverse genetics, performed on model organisms, to identify phenotypic effects of genomic interventions.
- X-chromosome inactivation analysis to study the possible effects on
  their offspring of particular patterns of X-mosaicism in females.

Molecular studies can, in principle, identify genes or alleles whose presence in
the genome plays a major causal role in the expression of a phenotypic trait.[18]
Such studies can distinguish among alternative hypotheses concerning distinct

---

18. For a good discussion of what and how genes can be said to cause or determine, see Waters 2004,
2006.

genetic configurations or between a hypothesis concerning genetic involvement and a null hypothesis. But they are not designed to distinguish among hypotheses about nongenetic causal factors or between genetic hypotheses and hypotheses concerning specific nongenetic factors.

•

Criticism of molecular behavioral genetics is directed less at its approach than at the assertion that anything particularly illuminating has emerged so far. Linkage analysis, claimed Billings, Beckwith, and Alper, has proven useful only for traits with simple Mendelian or X-linked modes of inheritance (e.g., the MAOA gene and aggressive behavior, or XQ28 and homosexuality among male relatives with a shared maternal line—both cases that hold only in limited populations).[19] Most behavioral traits will be both polygenic and multifactorial; thus, the association of any single gene with a behavioral phenomenon will be difficult to confirm, and nongenetic factors will also be involved. Initial reports of findings of genetic associations for bipolar disease, alcoholism, and schizophrenia have not been replicated. Indeed, replication of such results may be difficult given the small samples generally used. Each gene related to a polygenic trait may raise the probability of a given phenotype by only a slight amount, which would account for frequent failures of replication. Moreover, environmental factors within and outside the organism may play larger roles in the incidence of a phenotypic trait than genetic sequence alone. Genetic methodologies, designed to discriminate among intragenomic processes, and just possibly some intracellular processes involving genetic material, will not find those factors.

The small samples used in human studies are also problematic in that one or two misidentified alleles can affect the significance (positively or negatively) of a candidate association. Thus, even if there is a genetic component, it may be too fragile or too complex for current tools to isolate it.[20] These limitations have prompted researchers to adopt one of two strategies: (1) identification through the more powerful methods of GWAS of a set of genes that together account for a higher proportion of the distribution or (2) identification of environmental factors that, together with the gene in question, account for a higher proportion

---

19. Billings, Beckwith, and Alper 1992. On XQ28 and sexual orientation, see also King 1993 .

20. Thus, one response to the failures of replication reported in the Risch et al. meta-analysis (2009) is that the phenotype is not unambiguously operationalized or that further work to sort out genome-wide interactions is needed.

of the distribution than either genetic or environmental factors alone.[21] Since the results from genome-wide association studies show each associated gene raising the probability of a phenotypic condition slightly, with the whole set of identified genes, even then accounting for only a portion of the phenotypic variance thought to be genetic, the second strategy is likely to be more effective.

Any study associating a behavior with a genetic configuration must rely on assumptions about the base rate of the behavior in the population. The statistical significance of any association found via linkage or GWAS analysis is a function of the sample size and the base rate. The lower the base rate, the smaller the sample needed to support the claim that a given association is not a matter of chance. The higher the base rate, the larger the sample required. The Hamer study used a 2% base rate for the prevalence of homosexuality in the population. If the proportion of the population that qualifies as homosexual is greater—say, 4%, as some sociologists have claimed—the significance of the XQ28 association diminishes considerably.[22] A reliable estimate of base rate requires clear criteria of identification, and these same criteria must be used in any study assuming a given base rate. While there are estimates of base rates for many of the physical and mental disorders to which molecular methods are usually applied, such as susceptibilities to specific cancers or schizophrenia, I will argue in a later chapter that neither aggression nor sexual orientation is defined clearly enough to inspire confidence in any assumption of its base rate. Techniques for identifying genes and their allelic variants have consistently improved, but for most conditions of interest, it is still the case that allelic variation can be associated with only a small portion of the variance of that condition in a population. Assumptions regarding base rate and adequacy of sample size confer significance on individual studies but are shown to be problematic by subsequent studies that fail to replicate the purported association—for example, of given alleles with alcoholism or bipolar disorder. Polygenic traits are especially difficult to study, as single gene-behavior associations found in one study tend to be swamped by epistatic interactions in larger studies. The assertion that molecular (or traditional) genetic approaches using nonhuman experimental animals yield results that are relevant to humans must also assume that the behaviors studied are sufficiently analogous to human behaviors to warrant the inference.[23] Such claims must further assume that the

---

21. This strategy is the topic of chapter 6.

22. See references in chapter 3, note 17, for discussion of measurement of prevalence and for higher estimates.

23. For discussion and criticism of this assumption, see Schaffner 1998.

biochemical pathways that translate DNA sequences into RNA, thence to proteins, and thence to a series of higher-level phenotypes, from cells to tissues to organs, are also sufficiently analogous.

Conclusion

Molecular genetics promises to deliver more precise characterizations of the gene-behavior relation than is possible using the methods of classical population genetics. The cost of this precision so far, however, has been a reduction in the number of behavioral dispositions that can be attributed without qualification to genetic configurations. The power of DNA-centered research is irresistible, and molecular methods continue to advance. It is tempting to infer from the greater knowledge of genes and genetic processes that has resulted that genes are the fundamental causes of organismic processes, including behavior.[24] But this is to mistake relative quantity of information for completeness or quality of information. Molecular genetics methods are adequate to discriminate among alternative molecular genetic hypotheses, but they rely on assumptions about the base rate of a trait in the population, about the definability and adequate operationalization of the trait, and about the irrelevance of environmental variation to measuring correlation of genetic and behavioral variation or to comparative identification/measurement of potentially contributory alleles. Even the most powerful molecular methods account for only a portion of the variance of a trait that is thought to be genetic. These weaknesses have so far persisted, even as molecular methods have become more powerful. Molecular genetics has revealed the complexity of genomic processes, thus increasing our understanding of subcellular biology, but that sophistication is not yet matched by comparable progress in understanding their connection to processes at higher levels of organization.

---

24. See Waters (2007) for a careful account of just what causal inferences the new molecular research permits.

5

Neurobiological Approaches

N
europhysiological and neuroanatomical approaches to behavior seek to identify and characterize aspects of the neural substrate of behavior. Neurobiological research on humans rests on a vast literature detailing experiments with laboratory animals. Research identifying neural correlates of aggressive or sexual behavior in species ranging from rodents to primates has supported hypotheses of similar neural involvement in human behavior and stimulated research on the particular hormones, neurotransmitters, brain regions, and metabolic processes identified in the animal investigations. In the United States, the Decade of the Brain initiative (1990–1999) encouraged and funded a great deal of research on the human brain and nervous system. Earlier research (in the 1970s and 1980s) on living humans relied on so-called "experiments of nature," individuals exposed in utero to nonstandard levels of gonadal hormones, or victims of brain injury resulting from accident or stroke. Newer technologies permit study of living individuals whose behavior (or at least behavior proxies) and neural states can now be measured. Recent work on humans takes advantage of new imaging technologies such as PET (positron emission tomography) and fMRI (functional magnetic resonance imaging), as well as of new developments in neurochemistry and neuropharmacology. In addition, some studies continue to use traditional methods such as postmortem dissections for structural information.

Among the neurotransmitters, the role of serotonin in a variety of conditions and behaviors has received extensive attention.[1] By 1990 most researchers were in agreement that aspects of serotonin metabolism were involved in aggressive (or "impulsive-aggressive") behavior. Lower levels of serotonin and serotonin metabolites in cerebrospinal fluid were correlated with higher levels of aggressive behavior.[2] Research in the 1990s sought to elaborate the mechanisms of involvement and to separate out possible physiological confounders by zeroing in on aspects of serotonin metabolism. Serotonin diffused into the system can be absorbed by receptors before it reaches the synapses of the nerve cells whose activity it modulates ("presynaptic uptake") or absorbed by receptors on the membrane of those cells ("postsynaptic uptake"). There is a further process that terminates serotonin activity ("reuptake"). Given the multiple ways in which the amount of the transmitter reaching target neurons could be affected, was the culprit in the case of aggressive behavior decreased serotonin production or diminished uptake of serotonin by the appropriate receptors?

To address this question, a study by Emil Coccaro and colleagues investigated the possible involvement of serotonin receptors in the causal pathway.[3] The researchers administered a serotonin antagonist that would block the serotonin receptors to their ten subjects but not to the five controls. They then administered buspirone, an agent that physiologically mimics serotonin, to both subjects and controls. Receptor sensitivity was assessed by measuring levels of prolactin, a hormone released when serotonin or one of its agonists bind to serotonin receptors, before and after administration of buspirone. The idea was to identify the role of receptors by interrupting their function. Prolactin levels in subjects whose receptors were blocked were lower after buspirone administration (in relation to their individual baselines) than in controls whose receptors had not been blocked. Lower levels of prolactin, previously associated with higher levels of aggression/irritability, were established by this study to be a function of serotonin receptor (mal)function, rather than of volume of serotonin, a finding that implicated receptor function rather than serotonin production in serotonin's behavioral effects. Continuation of the research would involve investigating whether individuals identified as aggressive have fewer receptors or compro-

---

1. Other neurochemicals, such as norepinephrine and dopamine, also figure in 1990s brain and nervous system research.

2. This supported the focus on MAOA in the Brunner study, as discussed in the previous chapter.

3. Coccaro, Gabriel, and Siever 1990.

mised receptors in comparison with nonaggressive controls, which of the multiple types of serotonin receptor might be involved, and so on.[4]

Neurobiological research also examines structural correlates of behaviors of interest. Researchers can take advantage of the new neuroimaging strategies (PET scanning, magnetic resonance imaging) to find associations between loci of brain activity and violent behavior. For example, Adrian Raine and colleagues used PET scanning to study the brains of persons charged with murder who were referred to psychiatric clinics in preparation for insanity defenses or to determine competency to stand trial. They found diminished prefrontal cortical function in that group as compared with matched controls but similar levels of function in other brain areas in the two groups, leading them to conclude that prefrontal cortical dysfunction could be implicated in violence in some offenders.[5]

In 1989–1990 Simon LeVay conducted a comparative study of brain structure in heterosexual and homosexual patients who had died of AIDS.[6] LeVay chose to focus on the medial preoptic area of the anterior hypothalamus, which had been identified as involved in sexual behavior in rats. This area has four neuronal clusters, or nuclei, whose contribution to reproduction has been of interest to researchers. Comparing average sizes of the nuclei across (the cadavers of) homosexual men, presumed heterosexual men, and women undifferentiated with respect to sexual orientation, he found that one of the four nuclei (dubbed INAH 3) was, in the homosexual men's brains, closer in size to that of the women's brains than to that of the heterosexual men's brains. Just as the Hamer study, discussed in chapter 4, led to talk of a gay gene, LeVay's study led to talk of gay brains.[7] In 2001 William Byne and colleagues, using similar methods, replicated the LeVay study, finding that average size of INAH 3 in fourteen male homosexual brains was intermediate between that of thirty-four presumed heterosexual men and thirty-four presumed heterosexual women, and adding the information that only the volume of the nucleus varied, not the number of constituent cells.[8]

Work on laboratory animals involves manipulating levels of neuroactive sub-

---

4. See Olivier and van Oorschot 2005 for a review of serotonin receptor and aggression research in the intervening period.

5. Raine, Buchsbaum, et al. 1994.

6. LeVay 1991.

7. Indeed, LeVay and Hamer collaborated on an article (1994) in an effort to integrate their research findings.

8. Byne, Tobet, et al. 2001. Critics of both studies point to high standard deviations and to the role of outliers in producing the averages.

stances and observing the effects of those manipulations on behavior. Although most interest has shifted to neurotransmitters, neuropeptides, and other neurochemicals, testosterone was and still is quite intensively studied in this way. As understanding of this and other gonadal hormones has developed, it has become clear that the relation of circulating testosterone levels to aggressive (or any other) behavior is complicated. Animal research shows that testosterone levels are subject to experiential and environmental modulation, for example, so they cannot be considered an independent factor in the production of behavior.[9] Instead they are now understood as part of an interactive behavioral-physiological system. Human studies also show the complexity of testosterone activity, linking it to stress more closely than to aggression.[10] Nevertheless, some research continues to study the relation between prenatal exposures to "organizing" hormones and later behavior, and to make causal claims on their behalf.[11]

As an example of animal studies on the role of neurally active substances other than the gonadal hormones, the laboratory of Craig Ferris has been studying the role of the neuropeptide vasopressin and vasopressin antagonists in enhancing or decreasing the expression of highly stereotyped forms of aggressive behavior in several rodent species.[12] The value of stereotyped behaviors for research is that they can readily be identified and re-identified and that they are almost universal in a species.[13] Such behaviors as flank marking and assumption of typical postures for sex, then, are reliable indicators of some lower-level neural or physiological organization and thus useful for understanding the functions of substances like vasopressin (or serotonin, dopamine, etc.) in the organism. In addition to behavior, internal effects, such as binding to specific sites in the hypothalamus, are also studied. Thus, one might say, aggression is of secondary interest in this research—any easily observable behavioral effect would do as well. In the case of vasopressin, while its effects are uniform within a species, they vary across species. Moreover, as has been discovered for testosterone, vasopressin production is sensitive to experience.

---

9. See Wingfield (2005) for a brief recent review.

10. Archer 1991.

11. See Ehrenkranz, Bliss, and Sheard 1974 and Dabbs et al. 1995 for circulating testosterone; Hines et al. 2002; and Hines 2006 for organizing testosterone. For critical discussion of earlier work on the organizing role of gonadal hormones, see Longino 1990.

12. Ferris, Delville, Grzonska, et al. 1993; Ferris, Delville, Irwin, et al. 1994.

13. The dimorphism of some of these behaviors, that is, their differential distribution by sex, makes them particularly good objects of research, as their sex specificity provides some clues as to where to start looking for lower-level structures and processes.

Serotonin, the neurotransmitter central both to the Brunner MAOA study and to the Coccaro buspirone study and possibly the most intensely studied of these neurochemicals, was first studied in nonhuman animals: rodents and, later, rhesus monkeys. In both rodent and primate families, it was found to be inversely related to irritability—understood as a tendency to respond aggressively to adverse stimuli—which suggested a similar role in humans.[14] Accordingly, the role of the serotonergic system—the processes of serotonin release, transport, and reabsorption—in a variety of dysphoric conditions (including depression, aggression, antisocial personality, and addiction) has been studied via three primary methods in humans: (1) Concentrations of serotonin metabolites (e.g., 5-hydroxyindole acetic acid, 5 HT sulfate) can be measured in cerebrospinal fluid, with reduced concentrations of the metabolites signaling diminished serotonin activity. (2) The number of presynaptic receptor sites of serotonin reuptake can be indirectly measured (a greater number of sites indicates a higher rate of reuptake, leaving less serotonin available for binding to nerve cells and modulating their function). (3) "Pharmacochallenge" studies, which involve administering pharmaceutical agents, especially serotonin agonists or mimics (as in the Coccaro study), can be used to study serotonin activity in the brain, for example, the specificity of receptor sites or the role of serotonin reuptake in aggravating or diminishing aggressive responses.

All three methods of measuring serotonin activity are used in studies associating variation in serotonin activity with variation in human (impulsive) aggressive and antisocial behavior. The general idea is that lower levels of serotonin bound to nerve cells in the terminal areas of the serotonergic pathway result in higher levels of dysphoric conditions or of problematic behavior. The studies investigate variation in levels of serotonin activity in individuals with relevant diagnoses (under the *DSM* categories of impulsivity and antisocial personality disorder) or histories of criminal violence as compared with controls.[15] Advances in experimental strategies make possible the measurement and differentiation of responses in different chemical statuses within the same individual. In an effort to advance understanding of serotonin's mechanism of action, Emil Coccaro suggests that reduction in available serotonin, by whatever means, really affects the lower-level trait of impulsivity, which is expressed in aggressive behavior but

---

14. For an example, see Mejia et al. 2002.

15. Coccaro, Gabriel, and Siever 1990; Coccaro, Klar, and Siever 1994; Coccaro, Silverman, et al. 1994; Heiligenstein et al. 1992; Kavoussi and Coccaro 1993; Kavoussi, Liu, and Coccaro 1994; Lee and Coccaro 2001; Cherek et al. 2002; Olivier 2004.

also in risk-taking behaviors. Here, neurobiological work makes some contact with behavioral genetics work, where researchers have also advocated shifting the focus of attention from violence to impulsivity as a lower-level component of aggressive behavior more plausibly viewed as under genetic control.[16] The exact nature of serotonin's activity in the brain and nervous system is still not fully understood, but experimental demonstrations that decreased serotonin activity is associated with varieties of mental and emotional disorder, especially forms of depression, made the serotonin reuptake inhibitor fluoxetine—commercially known as Prozac—one of the best-selling drugs of the twentieth century (until overtaken by Viagra).

Neuroimaging has also made great strides since the investments of the Decade of the Brain. Computerized tomography (CT scanning) and nuclear magnetic resonance imagery (MRI or NMRI), especially functional MRI (fMRI), provide structural information to correlate with performance on various psychological tasks. Some of these tests (performed with either paper and pencil or touch screens) are surrogates for aggressive behavior itself. CT scanning uses X-rays passed through an individual's head from multiple positions to create a single three-dimensional image. MRI disturbs hydrogen ions with radio waves and employs a magnet to realign the ions. Active brain tissues manifest different resonance frequencies during the realignment phase than inactive ones, and these different frequencies are then used to create an image. The brain events that produce the image occur a fraction of a second after the brain events presumed to underlie the psychological events, but researchers take the imaged events to be residues and, hence, signs of those earlier brain events. Two additional techniques are used to measure brain activity. Regional cerebral blood flow (rCBF) studies use inhaled or injected xenon or similar tracers to detect areas of enhanced blood flow and, by inference, increased activity. Positron emission tomography (PET) assesses rates of glucose metabolism by means of a radioactive tracer that leaves a detectable residue at sites of metabolic activity.

Use of these imaging technologies is still costly, and the better the resolution the more so, thus limiting the sample sizes in these studies. And the data obtainable from the different types of imaging cannot always be correlated. Thus, any information gained by these strategies must be regarded as preliminary.[17]

---

16. Kreek et al. 2005; Coccaro, Bergeman, et al. 1997.

17. Issues similar to those studied by Rasmussen (1997) in the development of electron microscopy will undoubtedly arise with the future development of various visualizing technologies.

Nevertheless, researchers are attempting to identify abnormalities of function in particular brain areas with particular forms of aggressive or assaultive behavior. Given that the samples are small and constituted of individuals already suspected of mental illness or organic brain injury, their studies tend not to support generalization to violent or aggressive behavior in general. For example, since the point of the Raine study, mentioned above, was to find organic correlates of conditions indicating diminished capacity, the information it generated could be used to differentiate subcategories among violent individuals.[18] Even so, the results even for these purposes are preliminary, at best a basis for further research. Indeed Raine and colleagues have made suggestions as to how future research in this area ought to proceed.[19] Kent Kiehl and colleagues have also employed PET scans of the brains of incarcerated psychopaths (individuals convicted of particularly heinous crimes) to connect reduced or defective processing in the limbic area with heightened violence.[20] These studies correlate the verbal and brain responses to various images and compare the responses of the sample group with those of fellow inmates not labeled as psychopaths, as well as with nonincarcerated individuals.

In addition to such relatively direct observations of brain and neural function, a number of researchers have studied indirect signals of neural activity. One laboratory has investigated heart rate and heart rate differences among men who have engaged in spousal abuse.[21] Another has investigated heart rate differences between boys who persistently engage in fighting behavior (and have been identified by teachers and peers as disruptive) and controls.[22] These studies are best understood as developing heart rate measures as a research tool. That is, once such issues as stability over time, relation of resting to reactive heart rate, and the relation between heart rate and the social context and specific stimuli of an investigative situation are understood, heart rate will be usable as a signal that there is some organic dimension to whatever behavioral phenomenon is being investigated.

Both functional and structural studies have been performed in connection with sexual orientation. Well before the LeVay study, Gladue, Green, and Hell-

---

18. That such a warning is in order says much about the context in which research of this kind is received. However unrealistic, the hope is to find some genetic or organic defect that causes violent behavior.

19. Mills and Raine 1994; Raine, Brennan, and Farrington 1997; Raine 2008.

20. Kiehl 2006; Kiehl et al. 2001.

21. Gottman et al. 1995.

22. Kindlon et al. 1995.

man reported in the 1980s on a study of luteinizing hormone (LH) response to estrogen levels.[23] In women the monthly hormonal cycle includes a spike in LH levels following estrogen secretion. Gladue and his colleagues administered similar levels of estrogen to heterosexual and homosexual male subjects. They reported an LH response for homosexual men midway between the characteristic pattern for women and that recorded in their heterosexual male subjects.[24] The inference drawn was that homosexuality was hormonally influenced, if not hormonally driven. Meyer-Bahlburg and Dörner also investigated the possible role of gonadal hormones transmitted in utero on sexual orientation of offspring.[25] Ellis and Cole-Harding followed up these early studies in a survey of seventy-five hundred college students.[26] They found that the mothers of students identifying as homosexual reported higher levels of stress during the first trimester, a critical period during which gonadal hormones influence brain development. The connection is tenuous, as the actual in utero hormonal exposures of the students were not accessible, but it is consistent with the earlier hypothesis that maternal hormones affect the developing embryo, in particular with regards to development of sexual orientation.[27] One strategy employed by researchers to advance this line of research involves finding concordance of sexual orientation with traits generally thought to be hormonally determined. Handedness, relative finger lengths, and dermatoglyphics (finger print patterns) are among such traits. The idea behind these studies is that such concordance of patterns would suggest that the trait whose basis is unknown would have an origin similar to that of the trait whose basis is known. These studies have been performed both on gay men and on lesbians. Equivocal results have not discouraged researchers from continuing this line of investigation.[28]

•

The complexity of the human brain is an irresistible challenge to researchers. A number of prominent brain scientists have emigrated from other areas of the sciences (Francis Crick from physics, Gerald Edelman from immunology). But

23. Gladue, Green, and Hellman 1984.

24. For discussion, see Doell and Longino 1988.

25. Meyer-Bahlburg 1977; Dörner 1988.

26. Ellis and Cole-Harding 2001.

27. Further work is reported in Rahman and Wilson 2003.

28. See Hall and Kimura 1994; Williams et al. 2000; K. Cohen 2002; Mustanski, Bailey, and Kaspar 2002; Brown et al. 2002; Forastieri et al. 2003; Hall and Love 2003; Grimbos et al. 2010.

unraveling complexity is only one goal of brain researchers. Many have goals that are unabashedly pragmatic. Work on the serotonergic system, for example, has been accelerated by the aim of developing pharmacological therapies for dysphoric states, such as depression and anxiety. This is work that simultaneously advances the understanding of brain structure and function and of the role of neuroactive chemicals in regulating mood and behavior. It has obvious connections to the pharmaceutical industry, which sponsors some of the work and stands to benefit enormously from it.[29] It remains to be seen what the long-term effects of successful use of neuropharmaceuticals will be on research on brain structure and function, that is, what effects the successful regulation of neural states will have on the program of fully understanding them (e.g., understanding what proximate effects of the neuroactive substances give rise to the distal effects that are the object of control). The imaging work, locating brain regions that are active during, for example, erotic thoughts or hostile and aggressive thoughts, contributes to the larger project of mapping the brain. Eventually, combined with physiological research, it will give a fuller picture of the brain processes underlying cognitive, emotional, and behavioral phenomena.

Some of this work will undoubtedly lead to definitions and discrimination of components of and causal factors in behavior at lower levels of organization that can support a variety of applications.[30] For example, some research on neuroanatomical and neurophysiological correlates of violence may support the development of organic criteria for the application of the legal category of diminished capacity. This is clearly the concern of Raine and his colleagues and Kiehl and his colleagues.[31] Other work may lead to more finely oriented drug pharmaceutical interventions; Coccaro, for instance, in developing the view that impulsivity is the lower-level psychological-temperamental connection to a biological dimension of violent aggression, speculates that the palliative effects of fenfluoramine operate on impulsivity.[32]

Finally, the work associating neuroanatomical structures with sexual orientation purports to have the same social purpose ascribed to molecular genetic and behavioral genetic work, namely to normalize, by naturalizing, homosexual-

---

29. See Angell 2005; Krimsky 2003.

30. There are subtle questions here concerning the difference between causation, constitution, and correlation. See Churchland 1989; Searle 1992; Dennett 1991. Positions on these conceptual and philosophical issues play a role in the varying interpretations of what neurological research reveals about human cognition, affect, and action.

31. See notes 19 and 20 above.

32. Coccaro 1993; Coccaro, Silverman, et al. 1994.

ity. And like the genetic research, neurobiological research on behavior tends to transfer cognitive authority on behavioral matters to biomedical professions and away from nonmedical psychological or social service professions.

Methods, Scope, and Assumptions

The overall question addressed by the neurobiological approaches is *What role do neural structures and processes play in behavior B?* This question resolves into a variety of others:

- Can a specific, local, neural structure or process be associated with occurrences of B?
- Are the neural processes associated with B distributed or local?
- How are the processes associated with B activated or inhibited?
- How do the processes associated with B interact with other neural and organic processes?
- How can available techniques be improved to study the role in the expression of B of the neural processes associated with it?

The study methods employed to address these questions are retrospective, concurrent, and prospective. Retrospective methods include the use of autopsies to identify neurostructural correlates of behavioral patterns attributed to the individual and correlational studies of prison, clinic, and hospital records to identify associations between brain injuries or other trauma (e.g., birth complications) and later aggressive or criminal behavior. Here a typical hypothesis would be "Damage to region R of subcortical area S increases the frequency of aggressive episodes in patients characterized by conditions C." The contrast would be between subjects with damage to R and controls with no damage to R or with damage to other regions.

Concurrent methods include brain imaging or measuring changes in other physical parameters (heart rate) associable with exposure to particular cognitive or sensory stimuli. Here a typical hypothesis might be "Aggressive behavior is causally related to increased activity in region R of the neocortex." Contrasting hypotheses would concern lack of associable activity in region R, activity in other regions of the neocortex, or absence of effects of activity in R. Because imaging requires that the subject be immobilized, only thoughts of behavior can be correlated with measured neural activity. Thus such hypotheses would only be confirmed indirectly and against a background of assumptions regarding the connection of mental states experienced in an experimental setting and actual behavior.

Prospective methods include animal experimentation to identify the effects on behavior of organizational or activational exposure to bio- and psychoactive substances and clinical trials in humans to ascertain physiological, psychological, and behavioral effects of such substances. Here a typical hypothesis would be "Administration of P to individuals characterized by high incidence of B diminishes the incidence of B." Contrastive hypotheses would concern the effects of P on individuals not characterized by B, the behavior of individuals characterized by B who do not receive P, or comparisons of the effects of P with those of some other neuropharmaceutical also designed to increase serotonin availability.

While the retrospective and concurrent methods can answer questions about the strength of the correlation of a given neural structure, process, or type of damage with a given behavior, none can as yet produce data that definitively establishes that one kind of neural structure or process causally affects some behavior. While relevant to establishing causal claims, the correlations alone neither establish the direction of causality nor rule out a common cause. Prospective methods that involve manipulating a putative causal factor can establish that a given factor plays some causal role in diminishing, enhancing, or otherwise affecting a given behavior, but the knowledge thereby gained is as yet quite crude, as the mechanism of action needs to be identified by other methods and the effects of psychopharmaceuticals are generally quite a bit broader than the intended one. To summarize, then.

Retrospective methods include

- autopsy of individuals characterized by a particular behavior pattern to identify neurostructural correlates of those patterns;
- correlational studies of prison and hospital records to identify associations between brain injuries or other trauma (e.g., birth complications) and later criminal behavior; and
- correlational twin studies comparing frequency of neurosystem characteristics (e.g., serotonergic system function) in DZ and MZ twins.

Concurrent methods include

- brain imaging to identify areas of brain activity related to certain thoughts or sensory stimuli;

- measuring changes in other physical parameters (heart rate, salivary or serum testosterone) on exposure to certain cognitive or sensory stimuli, or in connection with certain kinds of experience; and
- measuring correlation of a behavior of interest with anatomical differences believed to have a physiological or genetic cause.

Prospective methods include

- animal experimentation to identify the effects on behavior of organizational or activational exposure to bio- and psychoactive substances, and
- clinical trials in humans to ascertain physiological, psychological, and behavioral effects of neuroactive substances.

Many of the studies on aggression work with human populations already clinically identified as antisocial, often prison inmates classified by the crimes of which they have been convicted or by psychiatric diagnoses or individuals under psychiatric care with a diagnosis of antisocial personality. When studying the behavioral effects of psychoactive agents, researchers also use categories of physical or verbal aggressiveness similar to those employed by the other approaches and ascertained via responses to questionnaires. In the case of sexual orientation, subjects are recruited through venues frequented by gay men or lesbians or publications geared to them. Sexual orientation is ascertained by self-report, or Kinsey measures based on questionnaire or interview responses may be used.

The point of studies of these kinds is better to understand the neural substrate of human behavior and action. In this regard they can be seen as part of the much larger research program of the neurosciences, but focused on particular neurotransmitters, structures, or other aspects of the brain and nervous system. The large gap between measurable changes in these features and changes in particular associated behaviors, however, means that this work is in its very early stages. Locally focused studies may reveal that a region or process is involved in a trait of interest, but they are not sufficient to show the full range of factors that interact with that process or region to produce the trait. Furthermore, while in some cases (e.g., injury that alters structure or function followed by a change in behavior) the direction of causality can be fairly confidently affirmed, in others the causal relations among empirically associated phenomena are still unknown. This is true of most of the neurobiological work on aggression and as well as of most of that on sexual orientation. Since in both cases some of the functional and structural differences could be consequences of behavior that has been canalized by other factors, their causal role cannot be unproblematically inferred.

The work done on aggression has a variety of pragmatic purposes. Research on serotonin, for example, informs psychopharmacological interventions, such as administration of Prozac or other agents to aggressive individuals. It thus is implicated in medical and in commercial networks, with corresponding social and political reverberations.[33] This work, as well as the neuroimaging and physiological research, also suggests the need to distinguish among subcategories of aggressive individuals. In some cases, a biological basis for a disposition to aggressive behavior can be identified, but it may be that there are several such bases. And there may be higher-level traits associable with some cases of aggressive disposition. Low impulse control, for example, crops up in a number of studies as a differentiating feature associated with various functional or structural differences. But not only do these intermediate physiological or psychological traits account for only a portion of the aggressive dispositions identifiable in any given sample, they are frequently also associable with behavioral or temperamental traits other than aggression.

Neurobiological research tends not to get caught up in debates as intense as those involving genetic and environmental researchers, although there are some disagreements as to whether neuroendocrinological or genetic explanations of sexual orientation are more promising.[34] Nevertheless, there have been criticisms of conclusions drawn either by researchers themselves or by scholars or media commentators. For example, the famous LeVay study on the correlation between sizes of hypothalamic nuclei and sexual orientation has been criticized for being based on too small a sample and for failing to rule out confounding factors.[35] Most of the neurobiological studies, like most of the molecular biology studies, are performed on small samples, and thus subject to similar limitations. The exception, the very large Ellis and Cole-Harding study, involved inference to neural development, not direct measurement.

Another interpretative challenge is posed by the difficulty of establishing the direction of causality or influence if a study can identify correlations only at a particular time. Thus, researchers have shown that testosterone levels increase *after* episodes of violence rather than before, as well as in response to stress, rais-

---

33. For example, some studies (e.g., Heiligenstein et al. 1992) received funding from the Eli Lilly pharmaceutical company.

34. See Bocklandt and Vilain 2007 for an argument that both aggression and sexual orientation are better explained by genetic than by endocrinological factors.

35. Fausto-Sterling 1992.

ing doubts about any general hypotheses concerning the causal influence of testosterone on aggression.[36] And because of their expense, many imaging studies are carried out on samples too small to support any causal conclusions. They are, furthermore, at a stage of correlating quite general patterns of thought with activation of brain areas and far from any kind of fine-grained analysis correlating specific content with specific neural patterns.[37] If one is a materialist or a physicalist, one would expect to find correlations of behavioral and mental phenomena with neural phenomena; the neural phenomena would be understood as constituting the mental phenomena, thus shifting the causal question from a focus on mental phenomena as explananda to the constitutive neural states. That is to say, the interpretive problems are not a function of metaphysical questions about the nature of mind or consciousness but concern inferences from the data to any causal hypothesis. Just what kind of correlation and constitution between neural and mental states is in question is a matter of ongoing conceptual and empirical investigation.[38]

A certain amount of this research is carried out to identify neuropathologies that might underlie otherwise puzzling violent or aggressive behavior. Such studies, especially those that establish a strong connection between a neurological abnormality and a behavioral one, are likely to be limited in applicability to specific pathological conditions. But aggressive behavior occurs in many contexts, not all of them expressive of pathology. Thus, this work, does not answer general questions about aggression. Something similar might be said about the research on sexual orientation that attempts to identify some specific neuroanatomical or neurophysiological condition with homosexuality. Where the model of inquiry is biomedical, involving a search for a specific neural abnormality that might underlie behavior, the explanatory scope is unlikely to extend beyond a small number of instances. This issue will be pursued further in chapter 6.

The methods employed in the neurobiological approaches involve a num-

---

36. Zitzmann and Nieschlag 2001; Sapolsky 2005; Booth, Granger, et al. 2006. Similar environmental and experiential sensitivity is demonstrated by other neuroactive chemicals. See, e.g., Huhman et al. 1991.

37. See, for examples of neuroimage-based correlations of behavior, mental states, and neural states, Greene, Somerville, et al. 2001; Greene, Nystrom, et al. 2004; McClure et al 2004; Sanfey et al. 2003. These studies seek to correlate observable neural phenomena with inferred emotional states and moves in experimental settings. The aim is to understand the interaction of brain areas associated with emotion or appetition with those associated with cognition; the specific behaviors involved in the studies function as indicators of mental states, rather than being themselves the object of the studies. An exception to the lack of specificity is the work of Suppes et al. (1996, 1999), who seek to associate distinct brain wave patterns with distinct sentences and concepts.

38. See references in notes 30, 37, and 41.

ber of assumptions. Neuroimaging techniques assume that brain areas showing greater glucose metabolism or differential disturbance of hydrogen ions during a particular thought process are causally involved in that thought process (and not epiphenomenally or incidentally affected by it). The assumption is that a brain area involved in the thought process will show activity. If additional conclusions are drawn about the involvement of that area in behaviors with which the thought process is concerned, then assumptions are also being made about the modality of the relation (causal or merely associational) between the process and behavior and, via the prior assumption, about the relation between the region of brain activity and the behavior in question. Thus, for example, studies that correlate areas of brain activity with hostile thoughts, and through those thoughts with aggressive behavior, may be described as finding associations or causal relations, depending on what assumptions are used in presenting the findings. Since the hope is to find causal relations, studies that find correlations are often interpreted as supporting causal relations, but the direction of causality is often only circumstantially supported. Furthermore, such lines of reasoning must suppose, first, that the thoughts can correctly be attributed to the participants and then that those thoughts are suitable proxies for the behaviors that are the ultimate object of interest. But the suggestion that impulsiveness is an intermediate trait suggests that thoughts attributable to a subject under observation by PET scanning, fMRI, or EEG cannot be easily equated with behavior.

The proposal that neural structures identified through autopsy are causally related to a behavior rests on the assumptions that the anatomical correlates have a function related to that behavior and that their development preceded rather than succeeded the relevant behaviors. To the extent that animal experimentation is relied on for support of these assumptions, both structural and functional cross-species uniformity and analogy of the human and animal behaviors is assumed. This assumption is not always satisfied. For example, the hypothalamic nuclei that in rats seem to influence choice of sexual mate are not the same as the ones in human hypothalami in which LeVay found size differences.

Clinical research that attempts to identify both physiological and behavioral effects of neuroactive pharmaceuticals need not make the same kinds of causal assumptions that the imaging and autopsy studies do. The interest here is in establishing that the administration of certain chemicals is followed by certain desirable states, and not by undesirable ones.[39] While some researchers are

---

39. This is the case with Heiligenstein et al. (1992), whose research was sponsored by Eli Lilly. For more documentation of the role of the pharmaceutical industry in research, see Angell 2005.

interested in showing the independence of particular brain-behavior relations from genes, others attempt to establish connections with heritability and genetic research and thus employ many of the assumptions of the behavioral genetics program.[40]

While much more is known about the brain and nervous system than was the case in the early 1990s, there is no overarching theory of brain function either to rally behind or to oppose. There are some general theories of brain function that drive experimental programs, such as the neuronal group selection theory advanced by Gerald Edelman, but the role of the brain in (human) conscious phenomena seems still as much a philosophical as a scientific question.[41] Research on neural correlates of aggression and of sexual orientation does not seem part of any specific theory of brain development and function but is a more piecemeal process, carried out under the not unreasonable assumption that accounts of brain and neural function will be part of any complete account of the mechanisms of behavior.

Conclusion

In human subjects, new imaging techniques permit identification of brain areas active during certain cognitive performances or while participants are subject to particular cognitive or emotional stimuli. Autopsies permit study of anatomical variation that may correlate with behavioral variation. Animal research permits broad-ranging investigation of neural processes, leading to specific hypotheses that can be followed up in human subjects. In humans, the role of neuroactive chemicals—neurotransmitters, neuropeptides, enzymes, hormones—can be studied by disrupting the processes of production, diffusion, uptake, and reuptake through administration of mimics and antagonists. Knowledge of brain structure and processes has advanced dramatically, but an understanding of their precise role in human behaviors still faces challenges. Advances in brain science, though extraordinary, have been piecemeal, focusing on careful measurement of chemical and electrical activity that can then be associated with higher-level behavioral (or temperamental) phenomena. Progress in identifying brain-behavior relata, as well as in understanding neurochemistry, indicates an

---

40. See, e.g., Alia-Klein et al. 2008, which suggests that the MAOA-aggression connection holds independently of genotype, and Coccaro, Bergeman, et al. 1997, which uses behavioral genetics methods to argue for a genetic basis to serotonin metabolism differences.

41. Cf. Johnson 1993; Edelman, Gall, and Cowan 1984. For more recent theoretical statements on the human brain and human consciousness, see Crick 1995; Damasio 1999; Edelman and Tononi 2000.

enormous complexity yet to be unraveled. In order to attribute causal influence on a given behavior to a given neural factor, studies must assume the independence of the factor studied from other factors and from the putative behavioral effect, and where variation in behavior cannot be fully accounted for by the factor under examination, must assume that the portion unaccounted for has independent explanations. There is no generally accepted big picture into which the empirical work can be set, and there remains a significant gap between it and the behaviors ultimately to be understood. The methods of identifying and measuring these behaviors are those already found to be common in the genetic and social-environmental approaches, thus, the neurobiological approaches are committed to similar assumptions concerning the individuation and operationalization of the behaviors of interest.

6

Integrative Approaches

W hile advocates of the approaches described so far all admit that biological and environmental factors interact to produce the behavioral outcomes they investigate, none, with the limited exception of classical behavioral geneticists, either pursues an interactionist program or considers how taking other factors into account might affect their approach, methods, and conclusions. There are, however, several research programs that do take understanding the multiplicity of factors and their interaction as the focus of their research. These include global integrationist views, such as developmental systems theory, and more local integrationist views, such as the GxExN integration approach developed by Avshalom Caspi and Terrie Moffitt and a multifactorial approach developed by Kenneth Kendler and colleagues.

Developmental Systems Theory

Overview

Developmental systems theory (DST) is the most ambitious of the integrative approaches, both in its view of interaction and in its intended explanatory scope. The approach has its theoretical base in embryology and developmental biology and was given its most thorough articulation by psychologist and developmental theorist Susan Oyama.[1] Here the central question is how the organism develops from a single fertilized cell into a mature individual characterized by multiple and specialized organs and tissues. Differentiation, the process of specialization of cells, is one of the key problems for developmental biologists. For the systems approach, genetic and environmental contributions to development are not separable; the relation between them is nonadditive and nonlinear. Nor can behavioral development be separated from the other dimensions of development. The object of inquiry is reconceived as a developmental system, which is a complex of interacting influences, some inside the organism's skin, some outside. Causation is understood as multiply contingent, and causal influences both select each other and determine each other's effects.[2] As Susan Oyama puts it, "change . . . is best thought of . . . as a system alteration jointly determined by contemporary influences and by the state of the system, which state represents the synthesis of earlier interactions."[3]

The late Gilbert Gottlieb, another of the principal researchers in this tradition, characterizes development as emergent, coactional, and hierarchical.[4] By "hierarchy," Gottlieb seems to mean the multilevel character of development and the interlevel as well as intralevel character of coaction; the term is not used to convey dominance or causal priority, but rather the unity of genetic, physiological, neurological, behavioral, and environmental aspects of the developmental system. It is used, therefore, much as it was by Grene and Eldredge, to represent something like degrees of embeddedness or of enclosure.[5] By "emergent," Gottlieb means that structural and functional complexity increases at all levels of organization of the individual as a result of interactions within and between these levels. By "coactional," he means that the factors involved in development interact. Not only do these factors not act independently, but they can also modify one another, thus altering their respective contributions in subsequent phases of development.

Developmental systems theorists describe the developmental process as "epigenetic." DNA is not static in the organism but conceptualized as the "effective genotype." Rather than think of the genes as stably active, DST sees the genome

1. Oyama 1985, 2000. Among philosophers, Paul Griffiths and Karola Stotz are noted expositors and advocates of the approach. See Griffiths and Gray 1994; Stotz 2006; Griffiths and Stotz 2006. Feminists have also embraced the approach. See Fujimura 2006.

2. Oyama 1985, 24.

3. Oyama 1985, 37.

4. Gottlieb 1991.

5. Eldredge and Grene 1992.

as consisting of potential patterns of gene activation that are differentially responsive to both intra- and extra-organismic environmental changes. As another researcher expressed the relevance of the concept of epigenesis to behavior, ontogeny (the development of the individual organism) is guided by three types of factor: (1) near universal maturational and contextual-relational factors, i.e., genetic and environmental features common to all members of a population or species; (2) idiosyncratic but stable maturational and contextual-relational factors, i.e., genetic and environmental features peculiar to the individual; and (3) constraints generated by the developing individual in the course of development.[6] Any individual developmental trajectory reflects a fusion of these three factors.

The empirical and experimental work of DST focuses on laboratory animals. These studies are intended to demonstrate either the canalization of behaviors by experience (rather than by genes) or the principle of coaction. An experiment with ducklings showed that they required in utero exposure to their own or their mothers' vocalizations in order to respond to species-specific calls after birth. An experiment with rat pups showed that pups bred to be spontaneously hypertensive (SHR) developed hypertension only if suckled by SHR mothers. They did not develop hypertension if suckled by non-SHR mothers, nor did non-SHR pups develop hypertension if suckled by SHR mothers; having been both birthed *and* suckled by an SHR mother was necessary. (Whether this reflects differences in behavioral interactions or substances in the milk was not, to my knowledge, examined.) Studies of behavioral plasticity in nonhuman animals are also cited as showing the complex coactional nature of development. Gottlieb and colleagues argue that an understanding of the complex relations among the factors influencing development is possible only if one conducts "controlled experiments that vary potentially influential factors in a systematic way."[7] Although experiments that would thus intervene on genes or physiological processes in humans are both unethical and impractical, the researchers claim that the similarity of molecular, physiological, neural, and cognitive processes between humans and selected nonhuman animals is such that some conclusions about developmental processes in general can be drawn. These are the claims about the epigenetic character of development just discussed.[8]

Another "systems" view is the dynamic systems approach. While often treated

---

6. Cairns 1991.

7. Wahlsten and Gottlieb 1997, 165.

8. DST theorists also stress their concept of a reaction-norm. See chapter 2, p. 35, for discussion.

as a version of developmental systems theory, on examination there appear crucial differences. Whereas *developmental* systems advocates stress the complexity of interaction of endogenous and exogenous factors in development, *dynamic* systems advocates stress the temporal and sequential nature of development—the dependence of one stage of development on earlier stages. They employ the apparatus of dynamic systems theory from the mathematical and physical sciences—in particular the notions of multiple possible steady states and of the importance of perturbations in shifting systems from one state to another—to study the development of specific capacities and behaviors. The late Esther Thelen was one of the pioneers of this approach.[9] She and her colleagues studied the development of basic motor skills, such as reaching and grasping, in infants, skills that involve cognitive, neural, and muscular coordination. They too conducted experiments "that vary potentially influential factors in a systematic way," but their interventions are limited to changing features external to the organism that do not have a permanent effect on it. Research by Thelen's group involves systematically varying elements of the situation implicated by the components of the skills studied, for example, changing the contents of the perceptual environment, altering the speed at which objects move in that environment, measuring arm angles and wrist torque, altering constraints on muscles (with weights), and repeating these variations over time to see how the organism learns to integrate component capacities into coherent behaviors. Motor skills develop as solutions to problems posed by the environment that stand in the way of attaining immediate goals, such as reaching and grasping a toy, and they develop in an order, from simpler to more complex. While the notion of the interaction of multiple factors is important to this approach, it is the temporality, the sequence of developmental events, that is the focus of attention.

Isabella Granic and her colleague A. K. Lamey have applied this interruption strategy to parent-child interactions in a laboratory setting.[10] They emphasize the abstract, mathematical nature of dynamic systems theory as a framework that can be applied to any number of types of concrete system, from physicochemical to behavioral. Their system of interest is the parent-child dyad, which they understand to manifest patterns of interaction that are differentially susceptible to settling into distinct stable states (attractors in the state space of pos-

9. See Thelen 2000; Smith et al. 1999.

10. Granic and Lamey 2002. For explicit development of the approach to understanding the development of antisocial behavior more generally, see Granic and Patterson 2006.

sible interaction patterns). One application of this model started by identifying two different types of parent-child dyad. In one, the children were diagnosed as "externalizing" (i.e., combative); in the other, the children were diagnosed as "mixed," both externalizing and "internalizing" (withdrawn, depressed). The experiment involved introducing a perturbation into the performance of a task and observing into what states the different dyad types subsequently settled. This is similar to the interruption of movement, changing of weights, and so on that Thelen describes in the development of control over limb movement in infants, but applied to interpersonal rather than intrapersonal processes.

The developmental systems approach is foursquare about the nature of developmental *processes* of individual organisms. This is different from the question asked by behavioral genetics, which concerns the extent of genetic influence on behavior. Behavioral genetics answers this by correlating genotype (or gene) frequencies in a population with behavioral variation in that population, or by associating individual genes or alleles with behaviors or traits partially constitutive of given behaviors. It is also different from the social-environmental question concerning the extent and nature of specific environmental influences on individual behavior. These programs both associate terminal points in a causal chain, while developmental and dynamic systems researchers emphasize the sequence of transformations between termini.[11]

•

The developmental systems approach is concerned with characterizing the full complement of interacting factors, the character of their interactions, and the sequence of stages in maturation. Developmental systems theorists see the point of their work as producing a comprehensive understanding of the mechanisms of development. The object of their interest is the individual organism as representative of the species, not populations. For them, the core questions are those of differentiation. In the realm of behavior, this means understanding how some behaviors are canalized, that is, fixed as stable habits or dispositions, while others remain malleable throughout an individual's lifetime. This insistence on variability and context dependence means that, if generalizations are to be obtained within this approach, they will not be generalizations about populations, indi-

---

11. And, it should be noted, the DST researcher would insist on the multifactorial character of the initial terminus. The ovum contributes cytoplasm and mitochondria in addition to nuclear genes to the blastula.

viduals, or their properties but generalizations about processes. Such generalizations will be more about abstractly conceived entities, such as canalized behavior, than particular behaviors, such as aggression or sexual orientation per se.

These researchers seem less concerned with practical applications than some discussed in earlier chapters, although some have argued that there are practical implications of taking the systems view. For example, developmental systems advocates have seen value in steering both researchers and clinicians away from overly simplistic conceptions of the causal relations involved in behavior. Lerner argued that a systems or contextual approach, because it involves a finer attention to differences among individuals and among environments, requires abandoning notions of the "generic child" (which tends to be a white, middle-class child) in the design of social intervention strategies.[12] Instead, researchers and clinicians should take into account contextual variability—both variation in developmental contexts and variation in the interactions of different individuals in the same context. Advocates, especially philosophical advocates, of developmental systems theory have presented the approach as an alternative to genetic determinism and as the source of a more adequate framing of evolutionary theory ("evo-devo"). Researchers, like Byne and Parsons, who advocate taking this approach to the development of sexual orientation see in it a way to naturalize homosexuality and bisexuality without locating etiology in a single factor.

While the dynamic systems approach to understanding the development of motor skills incorporates biological and environmental considerations, its approach to understanding the development of behaviors such as aggression or sexual orientation omits any explicit references to biological factors. To the extent its practitioners understand the individuals entering into dyadic or n-adic systems with relatively stable, interacting characteristics, the approach seems conceptually open to biological considerations. The goals articulated by Granic and her coworkers are to provide finer differentiations and classifications of the kinds of patterns that can characterize interactional systems and to develop and adapt the abstract dynamic systems approach to behavioral systems. Through information about the importance of interaction patterns in generating behavioral dispositions, researchers might identify strategies to alter those patterns in such a way as to generate or encourage the development of dispositions more conducive to the system's long-term welfare. Both systems views reject a conception of development as the rollout of a fixed program, whether encoded in the genes or elsewhere. Development, they emphasize, is contingent.

---

12. Lerner 1991.

Methods, Scope, and Assumptions

As in the other approaches, the overall question is *How does a capacity/disposition to B develop in individual organisms?* Application of developmental systems theory translates that into the following kinds of question:

- What developmental trajectories can be identified that culminate in B?
- What developmental factors (genetic, epistatic, intrauterine, physiological, physical, and social environment) interact in the development of B?
- Is the disposition to B canalized? If so, how?
- At what levels of organismic integration and organization do the causal/developmental processes relevant to B occur?
- Does the development of species-typical traits differ from the development of individually variable traits? (Of which type is B?)
- How do complexity of organization and specialization of function develop?
- How can intralevel and interlevel interactions be studied?

The dynamic systems approach adds several questions:

- What are the components of a given motor performance/skill?
- How does each of these components come to be expressed by the developing infant?
- What is the order in which they appear?
- How much variation in the number and order of appearance of components is compatible with the ultimate expression of the performance/skill?
- What are the patterns of interaction in a set S_1 of n-adic multiperson systems?
- Do the patterns in S_1 show systematic variation from those in set S_2?

In order to answer these questions the systems approaches employ methods in keeping with their overall theoretical frameworks. Developmental systems theory makes two claims that determine appropriate methodologies for answering these questions: (1) that the unit of analysis should be the (entire) developmental system, that is, the coevolving hierarchical set of interacting elements, and (2) that causal interactions are not just the combining of causal factors to jointly produce effects, but the mutual modification of such factors by one another. The theoretical claims of the dynamic systems approach are somewhat less abstract: the overarching claim is that mind and mentality (cognition and cognitive capacities) emerge through physical interactions with the physical environment. Its advocates' methods either presume or test these assumptions. One of the virtues of a dynamic systems approach in their view is its case-sensitivity, that is, the modeling variation admissible within the approach is relative to the particular case to which it is applied. Their focus is on particular behaviors, which they conceive in as complex a manner as the developmental systems theorists do the developmental system. They treat behavior as a dynamic system that exists in time, and the development of elements in a behavioral repertoire as phenomena to be studied as "processes in time rather than as static structures."[13] This applies both to individual behavioral repertoires and interactional behavioral systems.

Developmental systems researchers have concentrated their empirical work on animal models and on human infants and children, and thus have fewer empirical results related to behavioral dispositions manifest in adulthood or late adolescence than the other approaches surveyed. Nevertheless, individual researchers claim that this approach is the appropriate one to adopt either for human behavior generally or for some specific behavior.[14] Given the holism that characterizes this approach it is not clear how detachable or separable from the functioning of the entire human organism specific behaviors (like "starts fights without provocation" or "hits without provocation") would be or what strategies researchers would use to individuate and measure behaviors. In animals, by contrast, various behaviors amenable to experimental intervention have been studied. The main form of argumentation, however, seems to be either conceptual or addressed to the shortcomings of other approaches and reanalyses of their data. The experimental methods involve intervening in a developmental system to show that a given factor is (contrary to what might be expected from the perspective of another approach) essential to normal development of a trait or behavior. Experiments with animals involve breeding varieties of a species with common genetic profiles and then exposing those different profiles to different rearing conditions. When they are successful, such experiments demonstrate the necessity of *both* a specific genomic profile and a specific rearing condition for the development of a trait, whether that is a dysfunction like hypertension or a species-typical trait.

---

13. Thelen 2000, 7.
14. Byne and Parsons 1993.

Thelen and her colleagues have focused their studies on human infants and the development of specific motor skills, such as grasping, and cognitive-motor skills, such as performance in the A-not-B task.[15] Their studies systematically and sequentially vary features involved in the emergence of these skills, such as changing the degree of freedom of movement of the legs when learning to move the arms to achieve a certain goal, or adding weights to the arm. Applying this method to higher-level cognitively mediated behaviors would require similar variation (or its study when presented in nature) of the full range of genetic, physiological, and environmental factors involved in the fixation of particular behavioral dispositions. Its current application to such behaviors focuses on identifying personality types and interaction patterns by perturbing a relationship and observing differences in responses to the perturbation. These differences permit classification of interaction patterns. Independent identification of personality types through the use of standard inventories permits analysis of how combinations of particular personality types (which constitute systems) settle into particular patterns of interaction.

In contrast with the meticulous studies of human infants or interactional patterns in the dynamic systems approach, there are no human studies conducted explicitly using the developmental systems approach.[16] Its proponents' experimental and observational research instead focuses primarily on animal models and, in the case of fieldwork, animal populations. As applied to humans, the approach is used to provide an alternative interpretive framework for the behavioral genetic and social-environmental data that are already available and to serve as a platform from which to critique the interpretations given to these data by researchers committed to the respective approaches.[17]

Developmentalists are, like the environmentalists, accused of being nonscientific. Their alleged fault, however, is not social or political correctness but the reintroduction of vitalism.[18] Since the emphasis on epigenetic processes is vindicated by recent work on gene activation and regulation, one might see this as just metaphysics-mongering. The more substantive criticisms concern (1) the possibility of acquiring knowledge at all within the developmental approach and

---

15. This is a standard experimental test first conceived by Jean Piaget, in which an infant first learns the coordinated movements necessary to grasp B, and then is challenged to modify the learned motor pattern to grasp a differently placed item, A.

16. For further discussion of the differences between developmental systems theory and dynamic systems theory, see Keller 2005.

17. Wahlsten and Gottlieb 1997. See also the Stotz and Griffiths references in note 1.

18. Scarr 1993.

(2) the multilevel antireductionism of the developmentalists. Behavioral geneticist Sandra Scarr has been one of the fiercest critics of the developmental systems theory approach. In leveling the first of these charges, she argued that it is not possible to design studies that will identify causes of phenotypic outcomes if one is committed to treating all causes as mutually modifying one another and, hence, to not singling out any particular one for study.[19] She claims that the *how* question of the developmental systems approach will produce no or little knowledge in comparison with the *how much* question of the behavioral geneticists. There are two points that require disentangling here. The *how* question should be compared to the *how much* not as regards quantity, but as regards quality. Answering the *how* question results in different knowledge, some of it just not relevant to the *how much* question. Scarr is on better footing in her concerns about study design. While it is possible to create computer models of complex interactions, it is not possible to design empirical studies that can simultaneously measure the values of all the variables involved in a complex developmental process, as well as how they modify one another and result in some behavioral or dispositional outcome. Experiments are designed to keep all causal factors except the one or two whose action one wants to measure constant. And indeed, much of the animal experimentation of the developmental systems theorists involves demonstrating the necessity of some particular factor (or pair of factors) to a developmental process, not demonstrating the highly complex interactions the approach postulates. Thus, while an overall commitment to developmental systems theory can guide the design and interpretation of studies, the studies themselves cannot constitute a test of the approach, nor is it possible now even to conduct a study that fully exemplifies or fully reveals in an actual system the interactions posited by the theory.

The charge of antireductionism also requires untangling. Behavioral geneticists Burgess and Molenaar argue that reductionism is not incompatible with acknowledging multiple levels of organization.[20] They recognize multiple levels, which is one tenet of DST, but nevertheless claim that since the most "general," and hence most explanatorily basic, propositions will be found at the lowest level of organization, all explanation must ultimately invoke genes, and hence is reductionistic. If what they mean is that genes will be involved in any complete explanation, whatever else is invoked, then, since the developmental systems theorists would include genes among the relevant factors to be included in a

19. Scarr 1993.
20. Burgess and Molenaar 1995.

complete explanation, their approach would count as reductionist in that sense. There is another, and more common, sense of "reductionism," however, that holds that all causation takes place at the lowest level of organization. This is the sense rejected by DST. But they do not claim that genes are eliminable from all developmental explanations. Nor do they claim that all causation proceeds from some higher level of organization. Instead they claim parity for all levels of causal factor. This is the sense in which they would differ from Burgess and Molenaar, who privilege a "basic" level of causation and of explanation. However, there is a possibility intermediate between the reductionist view that holds all causation to be ultimately genetic and the parity of all factors in DST. Such an approach would allow for the multiplicity of causal factors without claiming that all constituent factors are equivalent.[21]

Some of the critical response to the developmentalists is defensive in character, addressed to the objections they raise to the behavioral geneticist program. Scarr claims that one doesn't need mechanistic knowledge, that is, knowledge of the process of development (or even of gene action) to identify causes of a phenomenon.[22] The *how much* question can generate causal knowledge of distributions of behavior. She also suggests that developmentalists are closet determinists. This is an odd accusation, given their insistence on the contingencies of interaction. Advocates of DST are concerned with acquiring causal knowledge of patterns of development in individual organisms, and they argue that genetic accounts of such development will always be incomplete, but a complete account need not be a deterministic account if the component causal factors have an irreducibly probabilistic relation to the outcome. The more temperate critics Turkheimer and Gottesman stress that behavioral geneticists and developmentalists are asking different questions: developmentalists are seeking to understand proximate mechanisms, while behavioral geneticists want to know about the variation in a population.[23] Furthermore, by focusing their criticisms only on single variable genetic analysis, developmentalists underestimate the kinds of information that behavioral geneticists can provide. One might add that, because advocates of DST are interested in the development of species-typical traits, they overlook the potential of behavioral genetics methods to parse influences

---

21. See Waters (2006, 2007) for development of such a position.

22. Scarr 1995.

23. Turkheimer and Gottesman 1991. This view is elaborated by philosopher James Tabery, whose work is discussed in chapter 8. If behavioral geneticists really limited themselves to such claims, it is hard to see how disputes could arise. The issues concern the inferences to genes and genetic causation and conflicting ideas of what the real questions are.

within a population in a uniform environment or to compare the extent of influence in different environments

Developmental systems theorists use the variability of heritability estimates when different populations in different environments are taken into account to motivate support for their overall approach. This argument assumes that interactivity of causes means that separation of causes is never possible. But careful experimental design can separate causes in a particular environment, and indeed, their own experiments require some isolation of potential causes in order to establish the interactions they hope to demonstrate. The error of those whom they criticize is in supposing that the knowledge produced under one set of experimental conditions can be exported without further investigation to different settings. The DST approach also assumes that humans and nonhumans are sufficiently similar with respect to overall developmental processes that conclusions based on animal experimentation will carry over to humans. Finally, their approach implies that to understand development it is essential to fully understand intra-individual processes. Their polemical writings also suggest that the only interesting biological question is a developmental question; the arguments rely on assertions that such and such a technique cannot illuminate development, that is, the processes whereby an individual of a given species comes to express a particular trait, whether eye color or a certain kind of intellectual ability. This is of course parallel, *mutatis mutandem*, to the criticism the behavioral geneticists make of them. If the point of such arguments is to delegitimate the use of behavioral genetics methods (e.g., twin studies) in the study of behavior, then they must assume that the only interesting question is this (narrowly defined) developmental one. But, as just noted, there are a number of questions behavioral geneticists can address that are not developmental.[24] The developmental systems approach has still to develop a full repertoire of empirical methods in order to develop data that can be used to support its proponents' particular theoretical proposals. Thus, many of the assumptions that would facilitate judgments of evidential relevance are not yet in play. Gottlieb's style suggests reliance on experimental reasoning, rather than on the collection of statistical data. The experiments described above are adequate to show the necessity of specific multiple factors in the development of some particular trait but, while showing *that* factors interact, are not powerful enough to support views about *how* they interact.

Although both systems approaches make global claims about development,

---

24. *Pace* Scarr's presidency in 1991 of the Society for Research in Child Development.

empirical demonstrations are restricted to the particular traits for which experiments have been designed. Until actually applied to the more complex activities and attitudes comprised in aggressive or sexual behavior, any inferential extension to such behaviors rests on assumptions of sufficient analogy between them and behaviors already investigated. In the case of the dynamic systems approach as applied to individual development, this would involve an assumption that basic motor patterns and skills such as grasping are analogous to aggressive behavior or to sexual/erotic orientations and behavior. Such an assumption would entail the supposition that the latter are as discrete and capable of individuation as the former.[25] It would also involve the hope that the components involved in the latter behaviors are as susceptible to study as are the components of the movements of infants. While the application of dynamic systems thinking to interactional patterns, as advocated by Granic and her coworkers, has an appropriate level of explanandum, it restricts the measured elements of the causal system to the psychological. In addition, the proliferation of psychological types and interactional patterns allows for multiple explanations suited to different types and patterns. The hope of a single explanatory scheme is thereby diminished. The dynamic systems approach does not make claims to completeness but is rather proposed as an approach for understanding the emergence of particular aspects of the organism. Thus, unlike the developmental systems approach, it is not proposed as a replacement for a genetic approach or for a social-environmental approach in the study of behavior.

GxExN

Overview

The GxExN program comes out of the work of Avshalom Caspi and Terrie Moffitt, whose investigations of the relations among early childhood experiences and specific genetic configurations have attracted a great deal of attention. Their first efforts were directed to exploring interactions among genetic and environmental profiles. They noted that the intersection of certain risk categories for a condition or behavioral effect showed a higher incidence of the condition/

---

25. One could agree, for example, with the motivating claim of the dynamic systems approach that mentality emerges and develops through bodily activity, and nevertheless reject the analogy by holding that the mental states involved in aggression or sexuality are themselves untethered from particular bodily movements and, hence, that the behaviors of which they are components are not subject to comparable categorization. See chapter 9 for further discussion.

profile than either risk category taken singly. This led them to develop a model of gene-by-environment (GxE) interaction in the development of behavioral/temperamental dispositions. Their strategy is to take some problematic behavioral profile X (aggressiveness, depression, antisocial behavior) and compare four subsamples—individuals with neither childhood abuse nor the genetic mutation, those with the mutation alone, those with abuse alone, and those with both abuse and the mutation. In many cases they showed that the group with both "risk factors" has a much greater incidence of X than the groups with only one, which showed either slightly elevated incidences compared with those with neither or no difference.

The basis of much of their work is a longitudinal study of 1,037 well-characterized New Zealanders born in 1971 and living in or near the city of Dunedin. One study centers on the relation between abuse as a child and a polymorphism in the promoter region of the MAOA gene in later (adolescent and adult) antisocial behavior in males.[26] The researchers compared the study population along several different measures: childhood experience (not maltreated, 64%; probably maltreated, 28%; severely maltreated, 8%), genotype (promoter associated with low MAOA activity, 37%; with high MAOA activity, 63%),[27] and several types of antisocial behavior. The idea behind using multiple types of antisocial behavior (including conduct disorder, conviction of violent offending, disposition to violence, and antisocial personality) was to establish the robustness of any result across different measures of aggression/antisocial behavior. Analysis of the data showed that genotype alone did not differentiate among the subjects with respect to the measures of antisociality. Maltreatment, on the other hand, did show a significant effect. And within the maltreated group, low MAOA was more highly correlated with antisocial behavior than was high MAOA. Whereas MAOA levels alone did not differentiate individuals classifiable as antisocial from those not so classifiable, within the group of severely maltreated individuals ($n = 33$), those with the low MAOA activity genotype were twice as likely to have been diagnosed with conduct disorder by age eighteen and 50% more likely to have been convicted of a violent offense by age twenty-six as those with the high MAOA activity genotype. This research provides an

---

26. Caspi, McClay, et al. 2002.

27. They make the assumption that the variation in the gene is effectively equivalent to the activity level of the enzyme. Alia-Klein et al. (2008) directly measured brain MAOA levels via PET scanning, in an otherwise similar study, and found an association of MAOA levels interacting with childhood maltreatment to increase scores on antisocial personality measures ("trait aggression"). The actual MAOA levels were not correlated with the genetic variation.

example of the interaction of genes and environment, as neither alone is sufficient for the outcome, while the group characterized by both factors had a much higher probability of committing serious offenses.

Since this early collaborative paper, Caspi and Moffitt have applied their integrative approach to a number of conditions, including depression, shyness, and cannabis use.[28] In a more recent paper, they advocate an integrated genetic-neuro-environmental approach (GxExN) for the study of psychiatric disorders. According to this more complex tripartite model, neurobiological structures and processes directly underlie specific psychiatric disorders; genetic factors determine those neurobiological structures; and environmental factors, such as divorce, death of a parent, spouse, or child, or loss of a job, differentially activate those neurobiological structures.[29] Thus, genetic and environmental factors interact at the neurobiological level, with genetically determined neurobiological structures constituting a vulnerability to environmental stressors, which is expressed in diagnosable psychiatric disorders.

The GxExN approach, then, has a narrower scope than that of developmental systems theory. While similar to DST in embracing the complexity of developmental processes, this model proposes a specific and measurable set of interactions to investigate psychiatric disorders. Rather than aspiring to a general account of the development of organisms, its advocates focus on offering causal accounts of specific psychiatric disorders by identifying underlying neural deficits caused by environmental stressors acting on genetically determined brain structures. Their goal is to provide an account of the development of (at least some of) these psychiatric disorders that permits effective intervention, either to alleviate their effects or to prevent their occurrence in the first place.

Methods, Scope, and Assumptions

While the GxExN approach might be understood as offering a general account of behavioral development, its goals specify a narrower set of questions:

- What disorder D_N of the neural substrate N is involved in the psychiatric disorder D_p?
- What effect do the identified environmental risk factors/causes, E, of D_p have on variation in N?

---

28. Caspi, Sugden, et al. 2003.
29. Caspi and Moffitt 2006.

- What effect does the identified genetic factor, G, in D_p have on variation in N?
- How do G and E interact in inducing the specific neural disorder D_N involved in the psychiatric disorder D_p?

To answer these questions and advance their integrationist program, Caspi and Moffitt have outlined six investigative strategies. Some of these amount to methods, while others are components of methods or proposals to acquire data in forms that make them relevant to the GxExN program:

1. Find or develop animal models of environmental pathogen exposure.
2. Given a characterization of the genotypic variation in a population, compare responses to environmental pathogens of the different genotype groups in that population.
3. Given a characterization of the exposures to a specific environmental pathogen in a population, study genetic variation among subjects exposed to the pathogen, especially with respect to the development of psychiatric disorders.
4. Include neuroscience measurements in epidemiological cohort studies of conditions such as depression, antisocial personality, etc.
5. Study the role of gene systems and sets of genotype polymorphisms in the pathology of psychiatric disorders.
6. Develop information on the relation of demographic variables such as age and sex to vulnerability to given psychiatric disorders.

These methods or components thereof employ elements from various single-factor approaches, include both prospective and retrospective methods, and specify ways to develop information about the interrelations of the factors that Caspi and Moffitt propose interact in the production of particular psychiatric disorders. The strategies they outline, thus, presuppose the success of methods already discussed in previous chapters, as applied to discrete psychiatric disorders. Their own work so far has consisted primarily in combining work from behavioral genetics and social-environmental research as it could be applied to the Dunedin sample, mostly along the lines suggested by strategies 2 and 3. For example, they have taken groups with a genetic profile associated with a condition and broken those groups into subgroups with and without experience of a particular life trauma, such as loss of a life partner, to see if the correlations within the genetically at-risk group vary by subgroup. In this, they have concentrated on genes associated with various components of the serotonergic system. Strategy 3 starts with groups of individuals who have experienced a particular life trauma. Strategy 5 would involve identifying individuals by genotype at birth and following their careers to see what behavioral profiles are associable with the different genotypes.

The Dunedin study involved post hoc genotyping for alleles independently established as associated with variation in the behaviors of interest. Given privacy, ethical, and scale concerns, the kind of full-scale prospective research required for an ideal implementation of strategy 2 remains out of reach, making identification of animal models, as suggested in 1, of crucial importance to the program. At this stage, Caspi and Moffitt have not incorporated explicit neurobiological research (strategy 4) into their work. And indeed, much work on the neuroscience side remains to be done, especially in the attempt to identify neural correlates of genetic or environmental risk factors, and that too will be advanced if appropriate animal models are found. Some research, of course, already exists, especially on the serotonergic system, for which the role of various genes involved in the production of monoamine oxidase (MAOA), which is in turn involved in serotonin metabolism, has been studied.[30] And there have been efforts to demonstrate the heritability of impulsivity under the presupposition of a connection between impulsivity and serotonin metabolism.[31]

Partly because Caspi and Moffitt's work is recent, and partly because it incorporates the work of genetic and social-environmental approaches, rather than setting itself up in opposition to either of them, it has remained somewhat immune to the critical animus characteristic of the nature-nurture debates. Nevertheless, to the extent that it relies on work from those approaches, it is subject to the same kinds of criticisms as they are. One study, by Neil Risch and colleagues, has taken explicit aim at the genetic research that supports Caspi and Moffitt's integrationist picture of depression.[32] The Risch group's meta-analysis of studies of depression and 5-HTTLPR (the gene implicated by Caspi and Moffitt)—prompted by a major conference called by the National Institute for Mental Health to discuss the GxExN approach to psychiatric disorders—found insufficient support for the claimed association between the gene and the condition. Some studies do indicate an association that reaches statistical significance, but as many or more are negative. Since the gene-disorder association was the basis for singling out

---

30. See references throughout chapter 4.

31. Coccaro, Silverman, et al. 1994; Coccaro, Bergeman, et al. 1997.

32. Risch et al. 2009.

the subgroup of individuals carrying the 5-HTTLPR serotonin transporter gene from among those experiencing stressful life events, if the association is spurious, the integrationist claim about depression collapses. While the meta-analysis calls the approach to depression into question, it does not necessarily affect the entire GxExN program. Similar meta-analyses with similar results on other gene-behavior/disorder associations would, however, cast serious doubt on the ability of the program to make good on its ambitions. Further blunting the critique, if the heightened genetic risk shows up only among individuals with experience of maltreatment or an environmental stressor, it might well not show up in the studies of a downstream condition in the general population.

The Risch meta-analysis is not the only empirical challenge to the GxExN model. Recent studies following up on the 1993 Brunner study have been equivocal: animal studies show elevated aggression associated with the "low" MAOA genotype,[33] but finer-grain human studies suggest involvement of other alleles and the considerable variation across ethnic groups in their frequency.[34] The study Alia-Klein and her colleagues conducted suggested to them that the low MAOA levels associable with aggression in humans are environmentally, not genetically, mediated.[35] While they did find an association between levels of MAOA measurable by PET scan and "trait aggression," their study failed to show any correlation between the measure of aggression used and MAOA levels, on the one hand, and genotype, on the other.[36] This study is even more challenging to the Caspi and Moffitt framework, as it suggests not that environment interacts with (genetically) fixed levels of MAOA, but that MAOA levels themselves fluctuate in response to environmental factors. In addition to these experimental challenges to the model, the need for suitable nonhuman analogues to fully carry out the program raises familiar questions about the extent to which nonhuman behavior can serve as a model of human behavior. Finally, given the role of behavioral pathologies in helping to identify specific neural or genetic correlates, as noted for the neurobiological research, the work is likely to be limited in its applicability to identifiable pathologies, rather than to behavior in general.

The GxExN approach requires some specific assumptions in its application to the particular traits for which it has been designed. These include assumptions that genes mediate the effect of environmental pathogens on disorders (through

---

33. Mejia et al. 2002.

34. Sabol, Hu, and Hamer 1998.

35. Alia-Klein et al. 2008. See also discussion in note 27.

36. See also Fowler et al. 2007.

their effects on brain structure and function), that experimental neuroscience can identify the neurological substrate of a disorder with sufficient precision, and that it is possible to overcome the challenge of small human sample sizes via idealizations, animal models, or even tissue stand-ins. With respect to the first, it could be that a common environmental factor (and not genes) induces nervous system vulnerability to what in that context becomes an environmental pathogen. Placing the focus on the neurobiological substrate, as neurophysiological approaches do, ought eventually to reveal this if it is the case. More to the point here, Caspi and Moffitt have articulated their model specifically as a model for psychiatric disorders, which are assumed to have clear diagnostic criteria and to be dysfunctional for the organism. And they treat psychiatric disorders not only as analogous to physiological disorders, but as stemming from physiological vulnerabilities, neural disorders. Their approach could be understood as an extension of an approach to the etiology of bodily dysfunctions that proposes that genes and environment act in a physiological substrate that is the more direct basis for manifestation of the dysfunction. As long as the psychiatric disorders investigated constitute discrete, identifiable syndromes, whether defined categorically or dimensionally, and have the causal structure proposed by the GxExN model, their approach will eventually (over many iterations) identify the etiologies of those syndromes. That is, as long as they can be understood on the model of specific physical traits like height or physiological dysfunctions like lymphoma, then the GxExN approach is likely to succeed with some pathological syndromes. Application of the model to behavior more generally, however, requires both the ability to associate nonpathological behaviors with noncompromised states of the brain and nervous system and, even more crucially, reliable criteria for the identification and individuation of behaviors. The challenges to meeting this requirement will be explored in chapter 9.

Multifactorial Path Analysis

Overview

The multifactorial approach employed by Kenneth Kendler and his collaborators attempts to quantify the roles of different factors, including genetic, social, experiential, and psychological, in producing a particular psychiatric outcome.[37] One of their studies employs data from a longitudinal study of 1,942 female

---

37. Kendler, Gardner, and Prescott 2002.

twins born in the state of Virginia between 1933 and 1972. Parents were also interviewed. Eighteen risk factors were sorted into five developmental periods in an effort to track the relative importance of genetic and environmental influences. Genetic risk was determined using standard twin-study methodology: concordance for depression in parents and offspring and in twins (correcting for dizygotic versus monozygotic status, but not comparing MZ with DZ). Other factors, determined through interviews, ranged from disturbed home environment in childhood to marital problems or other stressful events in the previous year. Interfactor correlations were assessed both diachronically and within the same developmental period, yielding a variety of information: that genetic risk alone was correlated with 10% of the variation in major depression in the sample during the year preceding the interview; that genetic risk slightly increased the risk of neuroticism, which in turn was correlated with 16% of the variation in major depression in the sample during the preceding year; and that the highest-valued risk factor (with a .35 correlation coefficient) was dependent on a major life-stress event stemming from the individual's own behavior in the previous year. The authors propose three major pathways through the risk factors: an internalizing pathway that starts with genetic risk, but for which early-onset anxiety also serves as an independent initiator; an externalizing pathway that starts with conduct disorder (which is itself correlated with factors in the same and the preceding developmental level); and a broad adversity/interpersonal-difficulty pathway that has disturbed family environment, childhood sexual abuse, and childhood parental loss as independent initiators.

Methods, Scope, and Assumptions

Kendler and his collaborators are frank in their discussion of the limitations of the study and of the ways in which various of the assumptions made or data-procurement methods used may have affected the final correlation measures. For example, they are not able clearly to sort genetic from environmental sources of risk in early childhood. Their aim is primarily to provide a method for displaying the multiple independent sources of adult depression, to show that individuals will vary in the pathways that culminate in depression and may display different combinations of risk factors. The value of this knowledge would be as a basis for intervention and prevention in the case of individuals who come to the attention of clinicians and display some combination of the identified risk factors. Adjusting the assumptions or introducing refinements into the data-collection process would result in adjustments of the values for correlations but would not provide grounds for reassessing the claim of multifactorial origin.

In a way this research represents a future for quantitative behavioral genetics. It employs some of the same methods, tracking concordance among biologically related individuals. It goes beyond behavioral genetics in identifying and measuring particular environmental conditions and events, rather than simply assessing environmental contribution by subtracting genetic variance from total variance or using other partitions of phenotypic variance. One feature that links this approach to classical quantitative behavioral genetics and differentiates it from the other two integrative approaches described in this chapter is its treatment of the relations among risk factors as additive and linear. Kendler's approach allows for complexity in acknowledging that a condition may have several independent origins, and it allows for interaction among them in that paths from different factors at earlier developmental stages may feed into another factor at a later stage that correlates more highly with the condition under investigation than any of the factors feeding into it. It must be stressed, however, that the interaction is statistical, in the sense of attempting to identify separate pathways that can, singly or in combination, result in the same outcome, rather than measuring the effects of risk factors one upon another to produce the outcome. It is a model for prediction. To the extent that there is an assumption that some of the events in the five life stages into which the measured risk factors fall are causally related, it is a causal model as well, but the authors make no effort to test causal hypotheses in the course of developing the model, nor do they make causal claims. To support causal hypotheses, at the least, longitudinal prospective studies would be required, following individuals, ideally genotypically individuated, from birth. Neither does the model identify the role of possible protective factors in mitigating the effects of risk factors. It is a map of risk factors and of the degrees to which they are correlated with one another and with the outcome of interest, depression. Finally, as is the case for all of the approaches surveyed, this approach assumes that the behavioral/temperamental outcome under investigation can be clearly and unambiguously defined.

Conclusion

Each of these types of integrative approach can be understood as an attempt to do justice to the multiplicity of empirical results achievable in the single-factor approaches and to their partiality and to integrate those various factors in ways suggested by the statistics of those results. Developmental systems theory, which holds that the unit of investigation is the entire developmental system, including factors internal and external to the organism, is the most ambitious and organically integrated of these. It is also the least susceptible to direct empirical validation, as that would require simultaneous measurement of an incredibly large number of parameters.[38]

The GxExN approach advocated by Avshalom Caspi, Terri Moffitt, and Michael Rutter focuses on the development of particular behavioral and temperamental pathologies. Like the developmental systems approach it incorporates the bodily and organic through its inclusion of putative neurosystem disorders in its etiological model, but its focus on disorders allows it to be more narrowly conceived.

The multifactorial approach attempts to quantify the impact of various measured factors on temperamental or behavioral outcomes, but apart from genes, these are environmental and psychological. The organic dimension remains an unmeasured substrate whose inclusion would undoubtedly affect the interpretation of the correlations presently ascertainable. To summarize, the more empirically tractable an integrative approach, the less explanatorily ambitious it is. Conversely, the more explanatorily ambitious, the less empirically tractable.

---

38. In this, DST faces challenges to those faced by climate modelers who work with complex models that are not directly testable. See chapter 7, note 6, for discussion.

7

Scope and Limits of the Approaches

T he previous chapters have surveyed five major approaches or families of approach to studying human behavior. These approaches are interested in proximal causes, that is, in factors that elicit or produce behavior (or variation in a behavior) in an organism in a population, rather than in ultimate causes, that is, factors that made particular behavior patterns adaptive in the course of the species' evolution or that account for the emergence of patterns in species whether or not they are adaptations. They have in common an interest in behavior as expressed by/in individual (human and nonhuman) organisms. But how behavior is understood as an object of investigation differs: it can be treated as disposition or as episode, as a dimension of variation in a population among populations, or as an individual characteristic. In some cases what is of interest is species-specific and species-typical behavior, in others, variation within species and populations.

Review of the Approaches

While the integrationist approaches have the ambition to offer a complete account of the development of some trait, the single-factor approaches focus on the role of some particular factor as cause or as focus of investigation. All of the approaches rely on assumptions, some required for application of a set of investigative methods to a phenomenon of interest, some to support extrapolations of results from particular studies or experiments to a broader population. Although conclusions are often hedged and results expressed as correlations or associations that meet some threshold of significance but require further research, the justification for pursuing any of these approaches is that it will reveal or assist in revealing significant causal information. Some assumptions are distinctive to a given approach; some are shared across approaches. In this chapter, I will summarize the results of the preceding five chapters, comparing the approaches with each other and with the less developed population/ecological approach, with regard to the kind of knowledge they can produce and the assumptions they must make. This will pave the way for more philosophical analysis in part 2.

•

Quantitative behavioral genetics. It might seem unfair or inaccurate to classify quantitative behavioral genetics (QBG) as a single-factor approach, given that it apportions variance between genes *and* environment. There are nevertheless two reasons to so classify it: (1) The practitioners of quantitative behavioral genetics are more interested (or have been more interested) in identifying the heritability of a given trait, and inferring the genetic contribution from heritability, than in identifying environmental contributions. (2) They employ one method, analysis of phenotypic variance by measured biological relatedness, to support their conclusions. The distinction between shared and nonshared environment is made by reference to phenotypic variance, not to differences in environmental factors themselves (e.g., between parental influence and peer influence, or media exposure and school, or socioeconomic status and intrafamilial processes). While critics emphasize the poverty of the approach's methods with respect to environmental parsings, it should also be noted that quantitative behavioral genetics does not (cannot) differentiate among genes either, except in cases of linkage, determined by the co-occurrence of two of more traits through multiple generations or matrilineality for X-linkage. Quantitative behavioral genetics is a practice of attributing portions of the variance in a phenotype in a population to genetic and environmental variance in that population, and doing so by using measures of biological relatedness among members of a sample of that population. It can discriminate among hypotheses that assign different heritabilities, understood as a ratio of genetic variation to total variation, to a trait in a population. Depending on the results of a study this can be useful in directing further research, whether into genetic bases of differentiation or environmental ones.

In humans the primary methods for sorting genetic contributions to variance from environmental contributions are twin and adoption studies. To support conclusions about a population under study, researchers must make several assumptions about environments and about their subjects. First, it must be assumed that the subjects do not constitute a biased sample. This may be difficult to substantiate for twin studies that recruit from twin associations or through media directed at twins, or for studies of sexual orientation that recruit from gay organizations or through gay media. Use of longitudinal studies following twins and adoptees from birth (or as close to birth as is feasible) and comparison of monozygotic with dizygotic twins (rather than twins with biologically unrelated individuals), when such comparisons are available, may overcome these problems but must still rely on assumptions about environments that cannot be empirically tested using the analytic methods of QBG. It must be assumed in the case of studies on adoptees that adoptive environments differ sufficiently from what would have been the environment of the biological family to support differentiations of biological from environmental similarities. This would require finding adoptees whose placement agencies do not have a policy of placing children in environments as similar as possible to what would have been their biological home environment and obtaining enough information about birth families to support the supposition that the birth and adoptive environments actually differ (and do so along relevant dimensions). In twin studies, it must be assumed that intrafamilial environments are sufficiently similar that differences in MZ/DZ concordance can be attributed entirely to biological variation between the two groups. When these assumptions are satisfied, QBG can give an estimate of the heritability of a trait in a given population characterized by a particular environmental configuration. To export a finding from a given sample population to the general population, it must be assumed that genetic and environmental configurations are distributed similarly across the populations. To use heritability estimates as the basis for claims about genetic causation, it must be further assumed both that gene-environment interaction is insignificant and that genes are the only vehicles for transgenerational trait transmission. Even if these assumptions are satisfied, the finding will only hold for that population or populations in identical environments. Finally, it must be assumed that whatever trait is under investigation constitutes a well-defined and re-identifiable phenomenon, that is, that the operationalizations and measures used in a study are representative of the trait more generally.

·

Social-environmental approaches. These approaches seek to identify determinants of behavior within the social environment. The methods employed involve correlation of traits of interest with environmental conditions or indices of environmental conditions (e.g., court histories of abuse), comparative observation of interactions of individuals in small groups, and observation of the effects of intervening on or coaching members of an individual's social environment on the individual's behavior. By comparing correlations, researchers can show that a given social-environmental factor is more highly correlated than another with a given behavioral pattern in the population under study. Interventions on elements of the social environment can support conclusions that the factor intervened on plays a causal role in the development or expression of the behavior. Social-environmental approaches discriminate among different environmental factors, rather than trying to show that environment rather than genes accounts for behavioral variation.

To argue from correlation to cause within the study population, researchers must assume that the correlated factor is independent of the behavioral trait and that there is not a common cause of both. This amounts to an assumption that all other possibly relevant factors (whether biological or social) are randomly enough distributed in the population under study, that only the effects of the factor under study will be revealed, or that the factor (or factors) that seems most highly correlated with the trait is not somehow elicited by the trait itself. As in the QBG approach, to extend results of correlational studies or intervention studies beyond the study population, researchers must assume that conditions are sufficiently similar across populations. This breaks down into two assumptions: (1) that conditions in other locations, for other populations, are sufficiently similar to those for the population under study, that is, that biologically relevant factors are similarly distributed and that there are no significant environmental additions or subtractions in those other populations, and (2) that efforts to reduce noise in laboratory or clinical studies haven't excluded factors that play a role in the world outside the study context. Moreover, it must be assumed that the trait under investigation is well defined and re-identifiable, with clear operationalizations and measurement standards, and that the trait studied in the sample population is the same as the trait researchers want to understand in the general population.

•

Molecular behavioral genetics. Molecular behavioral genetics (MBG) aims to identify specific genetic configurations that play a role in the development and expression of phenotypic traits. Until the advent of genome-wide association studies (GWAS), this was a bit like looking for a needle in a haystack, unless there were already clues to narrow the region of the genome in which one might find possibly relevant genes. Such clues were provided by X-linkage, by the association of a behavior with a physiological intermediary, some part of whose genetics were known (as in serotonin and the MAOA enzyme), or by the availability of multi-allelic markers. Once a gene region has been identified, animal researchers can engage in knockout experiments or other forms of reverse genetics to identify the role of a gene in some behavior. For human conditions, researchers are limited to identifying gene-behavior relationships through correlational studies that yield a statistical association between a gene (or gene region) and a behavior of interest. Most such associations of behaviors with single genes have correlated a gene with only a slight increase over the baseline in the general population, but GWAS has the potential to reveal the association of configurations of genes with behaviors. When successful, however, these studies reveal not *the* needle but many needles, and require further work to sort out whether multiple combinations of the identified genes are involved. MBG methods can discriminate among different molecular hypotheses within a particular level of analysis. As Kenneth Schaffner points out, this is just the beginning of an explanation of the phenotypic trait, as a complex series of physiological events and processes lies between genetic configuration and phenotypic expression.[1]

In assigning significance to any association, researchers must assume that they have correctly identified the base rate of the trait in the population, as base rate and size of the sample population determine the significance of any association. They must further assume that no other causal factors are influencing the observed correlation, that is, that the measurements made correctly reflect processes in the organism. To export any conclusions from the sample population, they must assume that general and sample populations are sufficiently similar genetically and environmentally. Finally, they must assume that the traits under investigation are well defined and operationalized and that the trait studied in the sample population is the same as the trait researchers want to understand in the general population.

•

Neurobiological approaches. Neurobiological research on behavior seeks to identify the neural structures and processes involved in behavior (both behavioral

---

1. Schaffner 1998.

dispositions and behavioral episodes). The methods, including postmortem examination of neural structures, study of brain damage, imaging technologies, biochemical analysis of neurotransmitters and neuroactive hormones, and analysis of physiological concomitants of types of experience, are each sufficient to discriminate hypotheses within the scope of what it measures. Autopsy can support conclusions that a given structure is or is not associated with some trait; imaging can support conclusions about increased or decreased activity in certain areas in connection with certain cognitive/emotive stimuli or performances; and so forth. Although animal research intervenes directly on neural processes, human research is limited to observing correlations between observable structures or processes as they occur in the organism and the phenomenon of interest, or to measuring the behavioral or temperamental effects of administration of psychoneuropharmaceuticals.

In reasoning from observed associations to causal conclusions, researchers must assume that the neural structures observed preceded the behavioral dispositions and that covariance between structures and dispositions indicates a functional role for those structures in the development and/or expression of those dispositions. In the case of imaging studies, researchers must assume that covariance between the brain activity imaged and a behavior or cognitive performance indicates a functional role for the activity, not an epiphenomenal relation to the behavior; the result of a common cause of the brain activity and cognitive performance; or even an effect of the mental activity. Researchers must also assume that the surrogates for behavior used in such studies (which require immobility) are highly enough correlated with the behavior of interest that they can be regarded as representative.

Pharmacological studies do constitute an intervention on the system, but conclusions about the substance or process acted on by the pharmaceutical and a behavior of interest in the sample population require ruling out placebo effects, and inferring to the general population requires assumptions about the activity of the pharmaceutical in the system, ruling out other biological factors, and assuming similarity in relevant respects of the sample and the general population. Finally, researchers must assume that the trait under investigation is well defined and operationalizable and that the operationalization in the study is distributed in the general population as it is in the study population.

•

Each of these four approaches provides at best a statistical association between a possible causal factor and the behavior or behavioral disposition of interest.

Thus, each identifies factors associated with only a portion of the phenomenon, whether the expression of the phenomenon in a population or the variance in expression of the phenomenon. There are two ways of thinking about this partiality. In one, we may suppose that the same phenomenon has multiple independent causes, each of which independently increases the probability of the phenomenon. This approach is linear and additive. In the other, we may suppose that the phenomenon is the outcome of interdependent causes, each of which (or some subset of which) is necessary but not sufficient for the phenomenon, and which individually have different effects depending on the configuration of the subsets in which they occur. The integrative approaches surveyed reflect these different ways of thinking.

•

Developmental systems theory/dynamic systems. These two approaches bring a systems understanding to thinking about organisms and development. Developmental systems theory (DST) focuses on the explanation of species-typical individual development. It is committed to viewing the organism as an element in a system, comprising a complex of interacting factors from genes through cells, through whole-organism phenotypes, to environmental factors playing a role in the system's sequential changes. Behaviors are often investigated as traits whose development illustrates the interactivity and interdependence of all these causal factors. The experimental work supporting this approach is performed primarily on nonhuman animals, and the application to human behaviors is an extrapolation by inference from the simpler to the more complex. The experimental and observational work in animals is generally able to demonstrate a gene-environment interaction for a given trait (and, in principle, two-way interactions for a variety of different factors), but not the full set of interactions proposed as constitutive of system change. The alternative hypotheses DST ought technically to be able to discriminate among would consist of different stipulations of the role of each of the different factors in producing some outcome, that is, alternative postulations of the system states that give way to the system state under study. As stressed in the previous chapter, however, it is not possible to obtain simultaneous measurements of all the factors of a system whose interaction determines the next state of the system. The approach, thus, still stands in need of investigative strategies that can reveal the actual causal interdependencies it postulates and discriminate between hypothesized interactions.

In insisting on the correctness of its approach, as against that of individual factor–focused approaches, DST assumes that the interactivity of causes means that it is never possible to study the causal influences of different factor types. Only a complete story will do, and the complex interactivity postulated means a complete story must incorporate all factors. In addition, DST polemical writings seem to assume that developmental questions are primary. For any extrapolation from an animal to a human behavior, more specific than indicating that multiple factors interact in the production of the phenomenon, it must be assumed that the animals provide adequate models of the human. There is undoubtedly a metaphysical sense in which this is (close to) the correct picture of organismic development: organisms are complex entities, highly robust within certain limits but vulnerable to extreme perturbations of the state of any of their components. The plausibility of the metaphysical picture is itself an inference from the demonstrated interdependence of genes and environment or physiology and environment in a few cases and the known partiality of any single-factor account. It serves a cautionary purpose, warning against overinvestment in any single-factor account. It also serves as a heuristic for researchers, encouraging investigation of possible interactions, even if the full picture can't be empirically demonstrated.

The dynamic systems approach is more limited in its explanatory ambitions. The systems investigated are much more local than those that are the concern of the developmental systems theorists. In one version, it focuses on the sequential and context-sensitive character of the development of motor skills, with a view to delineating the codevelopment of cognitive and physical elements of action. Each investigation focuses on the development of a particular skill, whether grasping or some locomotive ability. The muscular, perceptual, and conative elements involved can be thought of as a subsystem of the organism. This approach stresses the experience of the organism in the building-up of skills and in the development of behavioral dispositions. The experimental methods of the approach enable it to discriminate among hypotheses that assign different timing or order to the system elements. In the account of the development of antisocial or aggressive personality, the psychological orientation of subjects and interaction modes of n-adic groupings are stressed. Here the system is a social subsystem; a pair (or, in principle, a slightly larger collection) of individuals in interaction with each other is the object of investigation. No claim is made to comprehensiveness or comparative explanatory superiority, or to the independence of the studied factors from other factors (genes, prior experience). Since the central claim of the dynamic systems approach concerns the temporal and context-sensitive character of developmental processes, presumably whatever conditions/psychological orientations the individuals bring to an interaction would themselves be subject to the same explanatory strategy. At each stage, however, it is necessary to assume that factors other than the ones under study are randomly enough distributed as not to affect the outcomes under study. The dynamic systems account of antisocial personality, summarized in chapter 6, rests on work on small numbers of interaction pairs. Extrapolation to the general population would assume similarity in the relevant factors, and that the interactions studied in the sample populations are adequate models of the antisocial behavior that stands to be explained in the general population. This requires clear criteria of identification and re-identification, that is, that the behaviors be well defined and operationalized.

•

Gene x environment x neurosystem interaction. The GxExN approach utilizes research pursued in the single-factor approaches but, where both a genetic factor and an environmental factor are implicated in the expression of a trait in a population, inquires as to the frequency of the condition under investigation in that portion of the population characterized by both factors. That is, it breaks the populations studied into subpopulations characterized by just the genetic factor, by just the environmental factor, and by both. In a number of cases this strategy has resulted in finding that the subpopulation with the overlap of the two conditions shows a higher frequency of a certain adverse phenotype—whether antisocial personality, depression, or some form of substance addiction—than either of the single-factor subpopulations. This leads the researchers to postulate that genetic factors influence neural development in such a way that certain individuals characterized by a genetically based neural condition will be more susceptible to environmental "stressors" than individuals not so characterized. The alternative hypotheses among which the approach can discriminate would be hypotheses setting different frequencies for gene-environment overlap subpopulations than for non-overlap subpopulations. Identifying a neural basis for vulnerability would require neuroscience input into the program. Some such steps have been taken, as when researchers investigate the role of one of the MAOA genes in affecting patterns of serotonin metabolism or investigate the heritability of endophenotypes such as impulsivity. In the former case, a complication to this approach is presented by studies that indicate that the relevant variation that correlates with the phenotype is in actual MAOA levels, not variation in genotype.

The success of the approach requires satisfaction of several assumptions: first, that the methods of molecular behavioral genetics can identify genetic configurations associable with particular conditions or with an increase in their frequency and, second, that the methods of social environment–oriented research can identify environmental factors associable with increased frequencies of particular conditions. In addition, it must be assumed that experimental neuroscience can and will identify, through investigation of the populations characterized by an overlap of genetic and environmental vulnerabilities, the neural substrates of disorders selected. Full satisfaction of this assumption requires that adequate surrogates for humans exist and constitute suitable (morally, technically) objects of experimentation. Furthermore, the GxExN approach is developed specifically for conditions identified as disorders. Many disorders are the effect of a failure of one component (or a small number of components) among a large number involved in normal function. To apply the GxExN model to behaviors more generally would require that human behaviors be understood on the model of disorders.

•

Multifactorial Path Analysis. Here the aim is to identify the multiple factors involved in the manifestation of a disorder. The one exemplar of this approach described above involved identifying (or plausibly stipulating) developmental stages and relevant factors at each of those stages. Those factors are identified through methods appropriate to the associated single-factor approaches. Statistical analysis permits quantifying and comparing the degree of correlation of those factors with one another, within and across stages, and with the disorder under study. These methods could differentiate between hypotheses postulating different pathways through the complex of factors, pathways created by different correlation functions among the factors. Each one of those possible paths would identify a single factor or set of factors at each developmental stage, each of which is correlated with a factor or members of a set of factors in the stage preceding it. Through statistical analysis, each connection in a path can be assigned correlation coefficients that permit inferences as to which pathway is more prevalent in the sample, as well as permitting prediction of the probability of occurrence of a condition like depression on the basis of the presence of a factor or set of factors in an earlier developmental stage. On this basis, Kendler and associates claim that the adverse life-events pathway is the more common path to serious depression, without ruling out other paths. They emphasize that this is not a causal model, although it could serve as a starting point for causal investigations.

The quantification effort itself assumes that the relation between the factors is linear and additive, and that no unidentified factors are playing a role in the distribution of measured correlations. The full suite of correlated factors, as well as the outcome condition, must be well defined and operationalized. To extend a hypothesis beyond a study population assumes further that the relevant factors are distributed in the general population as they are in the study population.

Discussion

The so-called nature-nurture debate concerns which of these approaches is best. "Best" in this context can mean different things. It can mean most accurate. It can mean most fruitful. Or it can invite the question: Best for what purpose? The first of these is how the debate is understood by the general public and represented in general media: Am I shaped by my genes or by my upbringing? The second represents the issue in dispute among most researchers: Which approach will be most informative? Two features of these disputes seem most to animate the polemics with respect to these two perspectives.[2] One is the widely shared assumption that the approaches are all asking the same question. The other is the associated assumption that there is one correct way to represent the domain, the causal landscape, in and about which the question is asked. At some very general level, the various approaches *are* asking the same question: What causes/explains human behavior? But this question is both too broad and too vague to admit of any single answer. By "behavior," for example, do we mean tendencies in a population, particular episodes in the history of an individual, patterns of behavior, dispositions to respond to situations in one way rather than another, patterns of interaction? Do we mean behavior in general, or particular behaviors? What's more, to get a grip on this as a causal question, whatever the investigative approach, behaviors must be distinguished from one another and assigned unambiguous criteria of identification and strategies for determining when these are satisfied. This requirement presupposes that behaviors, at least those susceptible to causal explanation, constitute, if not natural kinds, phenomena at least stable enough to permit re-identification.[3] Causal questions may concern a population or species (Why do Xs Y?) or an individual (Why does/did X Y?), and may be evolutionary

---

2. Of course, the polemics are also fueled by the professional stakes at issue. These concern contests both for authority and for the grants and contracts needed to continue their work.

3. In his dissertation, Spencer (2009) introduces a notion of "genuine kinds," which are identified by sortal terms introduced and defined within a research program. The status of genuine kind is a minimum requirement behavioral kinds should meet.

(How did X come to be a Y'ing organism?), dispositional/episodic (Why do Xs Y when S?), or mechanical (How do Xs Y?). Particular sequences of movements are integrated differently into larger sequences of movement or into physiological sequences in different species. Thus, it is only meaningful to ask such questions relative to a particular behavior (or disposition) of a particular species, or an individual as a member of a species. Furthermore, some investigative questions are not directly about causes but about variation.

As I noted earlier, the question of causation has moved far beyond the simple nature-nurture dichotomy, as it is recognized that nature *and* nurture are causally implicated both in the evolution of particular behaviors and behavioral dispositions and in their expression in individuals and populations. And while the different approaches differ on the precise implications of this for research, they do agree that, for any causal factor, one can only think in terms of its contribution to (a) behavior relative to that of others. While a certain amount of popular press reporting is sensitive to this fact, it still presumes a context in which the first question, the question about accuracy, is preeminent. Reports thus tend to frame the issue as a competition: which approach is ahead in the race to understand a given behavior? The scientific disputes, however, are about the weight or strength of one type of factor vis-à-vis that of others, about the degree to which certain types of factor can be ignored, about the relative strength of investigative methods, about the relative value of different kinds of knowledge.

One additional point of comparison is worth remarking here. Popular representations of genetic research have tended to treat those approaches as more scientifically promising than social-environmental approaches. Even some of the researchers suggest that genetics will be more informative about the bases of behavior than environment, because environmental causes are so much more various and environmental influence so diffuse. This attitude may be a hangover of the "gene for" or "one gene, one disorder" idea, which suggested that for a given individuable behavioral phenomenon, where genes were implicated by quantitative genetics, it would be possible, via molecular techniques, to find a single gene. This attitude lives on in reporting such as that describing the low-MAOA genotype as "the warrior gene."[4] But what molecular genetics has taught us is that, while single genes in a given environment may increase the likelihood of expression of a behavior by some small amount, genetic influence is multi- rather than single-genic. Genetic influence is as diffuse in the genome as environmental influence is in the environment.

---

4. Gibbons 2004.

•

What the previous chapters have demonstrated is that all of these approaches are limited in different ways. Each has methods capable of discriminating among a distinctive range of hypotheses—with surprisingly little overlap. A study design that attempts to isolate one factor must employ methods that can measure it and the behavior under investigation. This requires stabilizing and neutralizing other contributory factors. Unless the factor is completely correlated with the behavior, as per standard empiricist categories of agreement, disagreement, and variation, the method can establish the necessity of the factor to production of the outcome, but not its sufficiency. Given the redundancy built into most complex organisms, even necessity may not be demonstrable. Instead, one can show that, *holding other factors/conditions constant*, a given factor is necessary. There are interconnections made across related approaches, as when quantitative behavioral genetics provides results concerning heritability that suggest molecular behavioral genetics methods have a chance at identifying a genetic configuration associable with a given behavior, or when physiological conditions associated with a certain behavior pattern can be associated with a genetic configuration. But these are piecemeal and subject to the same caveats regarding identification and measurement as have been articulated for the approaches themselves.

With the exception of quantitative behavioral genetics, which does attempt to parse the contributions made by genes and by environment to variation in a population, the methods available to these approaches are means of extending the approach to new phenomena or refining the account it makes possible. They are not methods for testing the approach itself against other approaches.[5] And even in the case of QBG, it is not testing the social-environmental approach against a genetic approach, or vice versa, but apportioning the variance in a population. Only a social-environmental approach that purported to make comparable claims about how much of the variance in a population could be attributed to environmental variation would be empirically competitive with QBG. Most social-environmental approaches, however, are interested in identifying social-environmental factors susceptible to intervention that would change behaviors and interactions, whether or not the behavior patterns are heritable.

Given the partiality of each single-factor approach, integrative approaches seem more promising as sources of causal understanding. But each of the three

---

5. See Longino (1990, 32–37) on goals of inquiry for more discussion of extending a theory or hypotheses versus testing it.

reviewed has limitations that preclude its being taken to offer any kind of comprehensive account. The developmental systems approach as yet lacks experimental methodologies that can test alternative hypotheses fully expressive of the interactions it postulates. The inherent challenges to simultaneous measurement of the multiple factors and their interactions suggest that this approach may never attain the status of an empirical science.[6] While there are some individual studies, like that of hypertensive mice, that illustrate the theme of coaction, they are mostly effective rhetorically, as rebuttals to excessive claims on the part of either behavioral geneticists or social environmentalists.[7] The metaphysical plausibility of this approach serves to remind us of the partiality of those approaches that are subject to empirical test, and it inspires researchers to design studies that might capture at least some of the complexity of developmental processes. Where the developmental systems theorists differ with the geneticists (or social environmentalists) is in their ambition to specify all of the factors involved in development and to characterize their interactions. To describe how a mechanism or process works in general is not to specify how it will operate in any individual case. Differences in initial conditions or in inputs to the system will change the subsequent interactions, and in some cases these differences will certainly occur below the threshold of detection. There may be multiple mechanisms leading to similar outcomes, and the interaction of the factors in any given mechanism means that specification of the parameters involved will support only probabilistically described outcomes.

GxExN seeks to reduce the uncertainties by targeting specific neural deficits, but the cost of this approach is restriction of the explanandum to clearly defined behavioral or psychological disorders. Results that hold for identifiable disorders like clinical depression, schizophrenia, or specific addictions are not exportable to nondisordered behaviors. The former have stereotyped modes of expression and (ideally) clear criteria of identification and measurement. The multifactorial path approach reaffirms the lesson that multiple factors are involved in the gen-

---

6. There are similarities here to the challenges of climate modeling. Part of the controversy about climate has to do with the different value placed on different kinds of evidence and the different strategies required to support model-based hypotheses, as distinct from measurement-based hypotheses. Earth's climate, however, is one (albeit huge) system, whereas in developmental systems theory we are speaking of millions of distinct systems, whose similarity is presumed but not demonstrated. Furthermore, climate researchers have observational and experimental resources not available to behavioral researchers. See Lloyd (2010) for a discussion of these issues in the context of climate research.

7. Some of the most persuasive research has been performed by West and King, but even they seem at best to claim that some given environmental factor must be taken into account, rather than giving a full account of the complex developmental processes. See West, King, and White 2003.

eration of conditions such as clinical depression and, depending on the correlations found, may also suggest that the GxExN approach should not hope to find one set of configurations that will hold for any given condition. But an approach limited in its ambition to finding correlation coefficients should not be expected to generate sweeping causal claims. At best, it can cast doubt on such claims by showing that the degree of correlation presupposed by some causal claim does not hold (whether generally or for some particular population). And even though it does not include all of the factors encompassed, in principle, by a developmental systems approach, it does suggest the magnitude of the measurement task.

Each of these approaches, then, is characterized by certain assumptions, some of which may be empirically confirmed or disconfirmed, others of which will resist such validation. Among the recurring assumptions are (1) that the approach in question has methods of measuring both the behavioral outcome that is the object of investigation and the factors whose association with it are the topic of investigation and (2) that the resulting measurements are exportable beyond the confines of the approach within which they are made. These assumptions will be the subjects of the next two chapters.

Analysis at a Different Scale: Human Ecology/Ethology

Left out of this analysis, so far, is another approach that draws not on the biology of individual organisms but on the comparative study of populations, whether ethology, which tends to focus on populations of conspecifics, or ecology, which focuses on the interactions of groups of organisms with each other and with their physical environment. One might call this approach human ecology (or human ethology), although this term has already been appropriated several times. While still marginal in the social sciences, human ecology is offered as a framework within which to investigate not just the impacts of human activity on environments but also the role of environmental and other system-wide variables on human social life.

There is some disagreement about the history of this approach. According to Mathias Gross, the expression "human ecology" was first introduced in 1908 in the collaboration of geographer J. Paul Goode and sociologist Edward Hayes.[8] According to William Catton the term was introduced in 1913 by Charles C. Adams in his *Introduction to Animal Ecology*.[9] The ideas of Goode and Hayes were

---

8. Gross 2004.
9. Catton 1994.

anticipated by Albion Small and George Vincent in the 1890s, reintroduced by Amos Hawley in the 1940s, and revived in the 1980s. Their 1980s revival occurred in the context of sociologists' response to the new environmental movements. More recently Elinor Ostrom has defended a similar approach, which she explicitly terms "ecological," under the banner of "institutional analysis."[10] Like the study of many other human phenomena, the study of these kinds of large-scale patterns and relations has bounced among disciplines and subdisciplines and suffered cycles of neglect and revival. Whatever its precise denomination or history, there does seem to be a recurring interest in frameworks of investigation that examine population-level phenomena such as differential distributions of behavior and the relation of these to other population-level variables. The approach has been extended to both of the families of behavior whose research has been the subject of the preceding chapters.[11]

•

As an example of what might be learned, consider the changes in primatologists' understandings of baboon social dynamics. Early study of baboons was focused on observation of savannah dwellers. These populations displayed high degrees of sexual dimorphism, of male dominance, of hierarchy, and of aggressive interactions among males. Notoriously, these behaviors were used by some primatologists and anthropologists as models of human behavior and social structure.[12] In the late 1980s researchers were able to study a population of forest-dwelling baboons genetically similar to the savannah baboons. Among these they observed less sexual dimorphism, less aggressivity, less male dominance, and less hierarchy than characterized the savannah dwellers.[13] Contrary to the prevailing view that held savannah baboon behavior to be genetically determined, their conclusion was that different settings characterized by different resources, different availability of resources, different vulnerability to predators, and so on, rather than genes, accounted for the difference in frequencies of behavior in the two genetically similar populations. While the question has not been definitively settled, the ecological explanation offers a robust alternative to the genetic one for differences in the behaviors of these populations.

---

10. Ostrom 2007; Brondizio, Ostrom, and Young 2009.
11. Philosopher Jonathan Kaplan (2000, 98–103) seems to embrace such an approach as an alternative to behavioral genetics but does not elaborate.
12. Weisstein 1971.
13. C. Anderson 1990, cited in Henzi and Barrett 2003.

Studies of human behavior have also employed this approach. Jeffrey Fagan studied rates of arrest and incarceration in New York City neighborhoods between 1985 and 1997.[14] He found, first of all, that these rates were highest in the poorest neighborhoods. He further found a disparity between the drop in crime rates in 1990–1996 and incarceration rates in the same period. Incarceration rates fell much more slowly than crime rates, and the excess incarceration was concentrated among nonwhite males living in the city's poorest neighborhoods.[15] Several factors contributed to this difference between changes in crime rates and in incarceration rates, including the narrowing of judicial discretion in sentencing and the reclassification of some misdemeanor drug offenses as felonies.[16] Given the brutalizing effects of incarceration, the lack of reentry programs for released convicts, and the high proportion of young men from the poorer neighborhoods who are in jail at any given time, Fagan proposes that incarceration simply begets more incarceration, that is, that in these neighborhoods, a flow in and out of incarceration has become established. The different structures of opportunity and constraint in New York City neighborhoods, including housing and employment availability, policing patterns, age structures, and so on, account for the differences in incarceration rates, not features internal to the individuals living in those neighborhoods. Whatever its merits, then, Fagan's study is not about why some individuals differ from other individuals in their manifestation of a particular behavior, but about distributions of interactions among the subpopulations (neighborhoods) of a given population (all of New York City) and their relation to the distribution of other population-level factors. Alfred Blumstein employs a wider lens on incarceration rates in the United States as a whole, puzzling over the disconnection between rising incarceration rates and flat crime rates and proposing various social structural explanations for further research.[17]

A population concept is also at work in approaches in the sociology of crime that focus on contextual factors related to crime rates in a region or area, rather than on factors peculiar to the perpetrators. Here too, different specific theories have been offered: the "broken windows" theory, which proposes that the

---

14. Fagan 2004.

15. Neighborhoods in which average annual household income was below $8,000 in some cases or $15,000 in others.

16. Fagan does not discuss the policies of lessened tolerance for minor offenses, to which lower crime rates are sometimes attributed and which could therefore be treated as accounting for both lower crime and higher incarceration.

17. See also Blumstein 1998.

accumulated uncorrected evidence of small, unpunished acts of vandalism or law-breaking morphs into higher rates of criminal activity, and the "collective efficacy" theory, which proposes that different rates of collective community engagement can account for different rates of criminal activity in different neighborhoods.[18] Whether the focus is incarceration or crime, these approaches emphasize not what makes the criminal engage in criminal acts, but the relation between the distribution and frequency of the phenomenon to be explained and the distribution and frequency of population-level properties.

In another example of a population approach, anthropologist Barry Adam reviewed a number of historical and anthropological studies of homosexuality.[19] He too proposed looking not at differences among individuals but at distributions of patterns of sexual behavior across societies differentiated historically and geographically. Distributions and frequencies of other-sex versus same-sex sexual interactions varied with variations in age and kinship structure, division of labor, and other population-level factors. In some societies, for example, there is a period of mandatory homosexual relationship for males as part of the process of gender differentiation. Others recognize a "third sex" comprising individuals who live in monogamous intimate relationships with individuals of the same sex. Adam proposes that these differences can be understood as a function of variation of population-level structures. Not only does this approach call into question the exportability of studies of the etiology of homosexuality in contemporary industrialized societies, it also raises questions about the very concept of sexual orientation they presuppose.[20]

While one might see this family of approaches as an extension of the social-environmental approach, there is a crucial difference. In the former, the focus of concern is the behavior of individuals and the factors that produce individual variation. Even in studies examining factors affecting interactions in *n*-adic groups, the explanandum is the behavior patterns of the constituent individuals.[21] In population studies, the focus is instead on the frequency and distribu-

---

18. For "broken windows," see Kelling and Cole 1996; for "collective efficacy," see Sampson, Raudenbush, and Earls 1997.

19. Adam 1985.

20. And it suggests interesting hypotheses about the increased visibility of gay- and lesbian-identified individuals in contemporary industrial and postindustrial societies. See Chiang 2009 for an effort to link distribution and frequency of homosexual activity to degree of urbanization.

21. To the extent that quantitative behavioral genetics purports to account for such distributions and frequencies, it might be understood as addressing a similar issue. However, comparative studies (e.g., Eley, Lichtenstein, and Stevenson 1999; Rhee and Waldman 2002) have as their problematic the question of

tion of behaviors or behavioral interactions in a population and their variation in frequency and distribution among populations. Thus the correlational/causal factors studied will also be population-level phenomena, such as age distribution, resource distribution, resource access distribution, income distribution, climate, and physical environment. The assumption of these approaches is that individuals resemble one another more than they differ in the (internal) factors influencing their behavior, and that differences in behavior are explicable by the structure of the choices before them, a structure that is externally determined.[22]

As with the social-environmental approaches, I am collecting multiple specific research approaches under one umbrella. Evolutionary theory, sociology, anthropology, and geography all offer instances of such studies. The umbrella covers such seemingly incompatible approaches as those that treat culture as a determinant versus those that treat social structures as a determinant versus those that treat natural phenomena as a determinant. While in principle recognizing this variety, my focus will be on the ways in which any of these approaches differs from those that focus on individuals and individual variation and that are most engaged with the nature-nurture debate. For my purposes, then, the difference is that between approaches that might take properties of populations in an aggregative manner, as the additive effect of properties of their constituent members, and those that treat properties of populations in a nonaggregative manner, as the effect of other population-level properties. These latter, ecological approaches offer different ways of thinking about causality and about classification, about what explains and what is explained in the sciences of behavior, thus enabling us to see the structure of the latter more clearly.

One goal of this approach or family of approaches is to identify factors, intervention on which might change the frequency or distribution of a behavior or behavioral pattern in a population, regardless of what causal factors influence its expression in individuals. By contrast, the prior approaches are suited to potentially identifying factors, intervention on which might change the frequency of a behavior in the population by changing the behavioral dispositions of individuals. It is useful to keep this contrast in mind as one contemplates the third question regarding which approach is best, namely "Best for what purpose?" It is this question that will be addressed in chapter 10.

---

whether heritabilities are the same or different in different societies or demographics, and do not address the differences in frequency of the behaviors studied in different societies.

22. The approaches must also assume that there is a way to identify and individuate populations that does not beg the question.

Part 2

Epistemological, Ontological, and Social Analysis

8

What We Could Know

In this chapter I consider what can be concluded from the preceding chapters about the reach of the scientific understanding of behavior. I extend the analysis of assumptions to underscore what the approaches have in common and what (really) differentiates them from one another. Since other philosophers have also treated these disputes, I consider some of their work in relation to findings about the structure of the research. Many of them follow the scientists themselves and attempt to show that one or another of the approaches is the correct (or incorrect) way to think about behavior, development, genetic and physiological factors, environmental factors, and the relations among them. I will argue, to the contrary, that each approach can provide partial understanding of human behavior broadly understood, and that each can provide answers to the questions specific to it, but that certain features of the investigative space preclude their full integration or the elimination of one or more in favor of a single encompassing and unified approach.

Integration or Differentiation?

The approaches discussed in part 1 of this book share common subject matters in the following sense. While aggression is multiply characterized, differently measured, and still, one might say, only vaguely defined, the multiplicity and differences in measure cross rather than differentiate the approaches. In the case of human sexual orientation, generally only one characterization and measure is used, the Kinsey scale. There are, however, debates about the adequacy of the characterization implied by the terms in this scale, and certainly questions about the criteria for placing individuals in any of the positions on the scale, given the extreme bimodality with which many studies operate.[1] The questions, methods, and assumptions differentiated in the previous chapters, however, constitute distinctive methodologies, or epistemologies, which, applied to the common subject matter, make salient different of its aspects, produce different causal histories, and generate different contextual linkages.

Such divergences generally produce one of two responses: efforts to reconcile the differences to provide a unified account or efforts to show that one or another is the correct approach. Philosophers and scientists both pursue these argumentative and analytic strategies. I argue instead that research pursued under the aegis of any of these approaches pushes them in nonreconcilable directions. Refining and improving the methods of a given approach enables researchers to produce better knowledge within that particular framework but does not produce tools for cross-approach empirical evaluation. Two kinds of assumption preclude such evaluation; one set has to do with the conception of causal relations employed, the other with the structure of the domain of investigation. I will first compare approaches that focus on variation among individuals, then move to a comparison of those approaches with approaches focused on variation among populations.

Assumptions about the structure of the domain have to do with the space of possible causes assumed by the approaches. Although researchers and most observers see the causal factors as existing in some kind of hierarchical relation, the logical relations among the approaches are better appreciated when no assumptions about their relative merits are made. A horizontal array that assigns each type of factor to a box, as in figure 1a, enables us to see how their respective roles are understood without incorporating assumptions about which are more basic. It can be expanded as additional factors are brought into the picture. Figure 1b, for example, adds a box for intracellular epigenetic factors, such as DNA involved in gene regulation or methylation rather than in protein production. Each approach parses this potential causal space in different ways: some attend only to factors in one block, treating the others as inactive; others try to assign portions of variance to different blocks, treating one subset as active and the

---

1. See, e.g., the papers in McWhirter, Sanders, and Reinisch 1990b. Issues concerning the characterization of common behavioral subject matters will be discussed in the following chapter.

Genotype 1	Genotype 2	Intrauterine environment	Physiology	Nonshared environment	Shared (intrafamily) environment	Socioeconomic status
(allele pairs)	(whole genome)		(hormone secretory patterns; neurotransmitter metabolism) **Anatomy** (brain structure)	(birth order; differential parental attention; peers)	(parental attitudes re discipline; communication styles; abusive/ nonabusive)	(parental income; level of education; race/ethnicity)

Fig. 1a. Potential causal space.

Genotype 1	Genotype 2	Intracellular epigenetic factors	Intrauterine environment	Physiology	Nonshared environment	Shared (intrafamily) environment	Socioeconomic status
(allele pairs)	(whole genome)			(hormone secretory patterns; neurotransmitter metabolism) **Anatomy** (brain structure)	(birth order; differential parental attention; peers)	(parental attitudes re discipline; communication styles; abusive/ nonabusive)	(parental income; level of education; race/ethnicity)

Fig. 1b. Potential causal space with extra divisions.

complementary subset as inactive. Thus, each approach measures variation in differently parsed causal space. These different parsings result in incommensurabilities among the approaches.

Quantitative behavioral genetics divides the space between genes and "shared family environment"—usually gross family characteristics such as socioeconomic status, since these are what are intended (and identifiable) as contrasts in twin and adoption studies (fig. 2). The variation of genes is studied against the macroenvironmental background, presumed to be stable, while presumed environmental variation is measured against presumed genetic similarity. "Nonshared environment"—for example, what must account for the portion of variance left over, which may include such phenomena as differential parental treatment of siblings—was early on treated as noise or as itself a genetic effect.[2] As we have seen, behavioral geneticists have begun to characterize nonshared environment as a distinct component in the causal mix, as it seems to account for a nontrivial portion of phenotypic variance.

Molecular genetics seeks to associate specific traits with specific alleles or

2. See Scarr 1992; Plomin, Owen, and McGuffin 1994.

Genotype 1	Genotype 2	Intrauterine environment	Physiology	Nonshared environment	Shared (intrafamily) environment	Socioeconomic status
(allele pairs)	(whole genome)		(hormone secretory patterns; neurotransmitter metabolism) Anatomy (brain structure)	(birth order; differential parental attention; peers)	(parental attitudes re discipline; communication styles; abusive/ nonabusive)	(parental income; level of education; race/ ethnicity)

Fig. 2. Causal space for quantitative behavioral genetics. Light gray = measurable space. Dark gray = unmeasured space.

Genotype 1	Genotype 2	Intrauterine environment	Physiology	Nonshared environment	Shared (intrafamily) environment	Socioeconomic status
(allele pairs)	(whole genome)		(hormone secretory patterns; neurotransmitter metabolism) Anatomy (brain structure)	(birth order; differential parental attention; peers)	(parental attitudes re discipline; communication styles; abusive/ nonabusive)	(parental income; level of education; race/ethnicity)

Fig. 3. Causal space for molecular behavioral genetics. Light gray = measurable space. Dark gray = unmeasured space.

allele combinations (fig. 3). Whether following up on linkage studies that suggest a particular section of the genome or on genome-wide association studies, molecular geneticists too treat the portion of variance not attributable to genes as environmental. Unlike quantitative behavioral genetics, however, molecular genetics can initiate investigations into the genetically influenced physiological or anatomical substructure of behavioral traits by focusing on the protein products of DNA-sequence activation.

Environmentally oriented researchers treat the space differently (fig. 4). They too divide it between the genetic and the environmental, but for them the genetic side is treated as generic or uniform and the environmental side as the location of effective variation, both in gross characteristics, such as family socioeconomic status, and in more finely grained ones, such as disciplinary practices, forms of endearment and other aspects of one-to-one interaction, maternal gender stereotypes and attitudes, and so on.

Comparison of the three arrays shows one source of incommensurability. For

Genotype 1	Genotype 2	Intrauterine environment	Physiology	Nonshared environment	Shared (intrafamily) environment	Socioeconomic status
(allele pairs)	(whole genome)		(hormone secretory patterns; neurotransmitter metabolism) Anatomy (brain structure)	(birth order; differential parental attention; peers)	(parental attitudes re discipline; communication styles; abusive/nonabusive)	(parental income; level of education; race/ethnicity)

Fig. 4. Causal space for social-environmental approaches. Light gray = measurable space. Dark gray = unmeasured space.

quantitative behavioral geneticists, intrauterine developmental factors of what-ever origin,[3] because they are not identified by the kinds of twin and adoption studies they can perform, count as environmental or are dismissed as noise. For environmental researchers, any factors, including intrauterine ones, that don't belong to the measurable environment are treated as inactive, even though some might well be properly classified as environmental. Specific distributions be-tween genetic and environmental causation will vary between the approaches, even though their main interests are distinguishing among causal contributions by the factors they study.

The neurobiologists can be seen as opening up the middle area between genes and environment by focusing on the organic—anatomical and physiolog-ical—substrate (fig. 5). While many neurobiologists may think that the neural substrate is genetically determined, and behavioral geneticists certainly act as though the organism inside its skin is a direct expression of the genome, suf-ficient experimental work exists on the plasticity of neural structures to make clear that such a supposition is just an assumption.[4] Thus, unless experimental data indicate otherwise (for example, the apparent relation between a mutation in the MAOA gene and aspects of serotonin metabolism), the organic, neural substrate must be treated as an independent factor relative to the organism's genotype. That is, in studying its effects, researchers must treat it in isolation

---

3. These can range from maternal mRNA that affects offspring phenotype to maternal stress or nutrition that affects fetal development.

4. Research showing the mutability of the rate of testosterone secretion in response to the same stimulus after changes in social or environmental situation demonstrates the complexity of the interactions of the various systems involved. Similar results are being obtained for the serotonergic system. See Yeh, Fricke, and Edwards 1996; Alia-Klein et al. 2008.

Genotype 1	Genotype 2	Intrauterine environment	Physiology	Nonshared environment	Shared (intrafamily) environment	Socioeconomic status
(allele pairs)	(whole genome)		(hormone secretory patterns; neurotransmitter metabolism) **Anatomy** (brain structure)	(birth order; differential parental attention; peers)	(parental attitudes re discipline; communication styles; abusive/ nonabusive)	(parental income; level of education; race/ethnicity)

Fig. 5. Causal space for physiological and anatomical research. Light gray = measurable active space. Medium gray = possible determinants of active space. Dark gray = unmeasured space (although some studies control for shared intra- and interfamily environments).

from the factors that produced it. But this will perforce ignore feedback effects that may reset or otherwise alter the system. Attempts to read significance into neuroanatomical correlates of aggression or of sexual orientation often treat the neural structures as potential causes rather than effects of such traits or as joint effects of a common cause. Thus, the active, measurable causal space of the neurobiological approach is occupied by neural structures, systems and processes, with genes, environment, and other developmental factors interacting in the background. The interest here is in the role and function of developed structures, not in the processes leading to their development.

For developmental systems theorists, all of these factors are potentially in the same causal space, as are the interactions among them (fig.6). Thus, fine-grained environmental factors (like birth order) will interact differently with one set of endogenous factors (e.g., a given pattern of endocrine secretion) than with another. And these interactions will themselves vary depending on other factors, such as cultural values or differential social reward systems. For developmental theorists, the environment in and with which genes interact includes the intracellular and extracellular physiological environment, in addition to the social and other factors external to the individual. This heterogeneity of the causal space prompts the behavioral geneticists' complaint that the developmental approach is methodologically unrealistic. How is it possible to develop a model of, say, intelligence differences or sociality differences if all those factors must be included? Indeed, developmental theorists may not be interested in modeling a single trait as opposed to articulating a general model of the interdependence of all the factors in the causal space. That is, they may be less interested in understanding particular behaviors than in understanding the matrix within which any behaviors

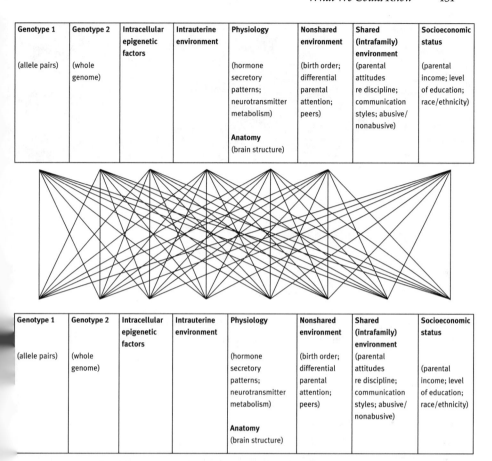

Fig. 6. Causal space for developmental systems theory. A partial representation of the causal relations. Each type of factor can affect each other type of factor and affect how each other type of factor affects higher-level organismic state.

develop. As noted above, to date its empirical program has consisted in demonstrating the insufficiency of single factors and supporting the conclusion that there must be multiple factors. It has not developed resources for studying the interaction (or coaction) of these factors empirically. And according to some of its critics, it is not even possible to do so. In any case, the DST approach ends up treating the cause of a change in the system as the prior state of the entire system and the effect of a change as the subsequent state of the entire system.

The GxExN approach utilizes ongoing research represented in several of the above grids. The restricted approach to factors, as well as the employment of research from extant single-factor approaches and the treatment of a particular

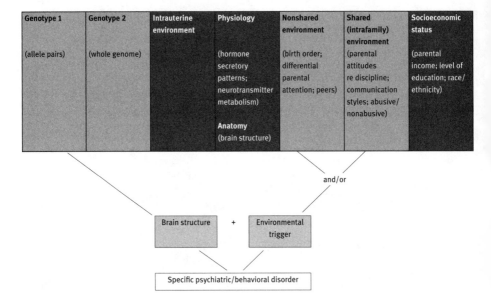

Genotype 1	Genotype 2	Intrauterine environment	Physiology	Nonshared environment	Shared (intrafamily) environment	Socioeconomic status
(allele pairs)	(whole genome)		(hormone secretory patterns; neurotransmitter metabolism)	(birth order; differential parental attention; peers)	(parental attitudes re discipline; communication styles; abusive/nonabusive)	(parental income; level of education; race/ethnicity)
			Anatomy (brain structure)			

Fig. 7. Causal space for GxExN approach. Light gray = postulated active space. Medium gray = measurable determinants of active space. Dark gray = unmeasured space.

disorder (rather than a comprehensive state of the entire system) as the effect, distinguishes it from the DST approach. GxExN proposes specific interactions among specified genetic configurations and environmental factors, such as the MAOA gene and experience of abuse, as causes of specific psychiatric or behavioral disorders. While certain parts of the array are silent as initiating causes, they reappear as effects and intermediate causes somewhat downstream (fig. 7).

•

Thus, even though there is, from one point of view, a common phenomenon to be understood—behavior, or rather, specific behaviors or behavioral patterns such as aggressivity or sexual orientation—each approach brings with it a prior and distinctive representation of the domain of investigation. Each employs distinctive strategies to sort alternative models or hypotheses regarding etiology. The result is a partial rather than a complete specification of alternatives, each specification incorporating partially overlapping but differently measured sets of phenomena. The partiality is not a simple part-whole relation, as the approaches can't simply be integrated into a single picture. Since the methodologies developed within the different approaches are designed to differentiate between causal influences in a particular space of possibilities (biologically transmitted factors versus shared family characteristics, one genotype versus another genotype, one social influence versus another social influence, one brain area versus another brain area), and since the spaces utilized, as well as their contents, differ from approach to approach, one approach does not have the capacity to discriminate among the causes proper to the other approaches (genotype G versus social experience S versus neurotransmitter N).

Consider what analysis of variance (ANOVA), the one method designed to differentiate among approaches, can and cannot tell us. Designed to apportion variance in a trait in a population to different factors, it is employed by quantitative behavioral geneticists to establish the heritability of a trait in a population but is, in principle, also available to researchers seeking to establish environmental etiology. Quantitative behavioral geneticists employ ANOVA in a way intended to separate out the proportion of variance in a trait that can be assigned to genetic variation (determined by finding degrees of concordance in expression of a trait among biologically related individuals). An environmentally oriented researcher could similarly employ ANOVA to separate out the proportion of variance in a trait that can be assigned to variation in a set of specific measurable environmental factors. As I have been suggesting, the different foci of the two approaches means that they will categorize potential correlates of observed variation differently. A factor classified as nongenetic and hence environmental by a genetically oriented approach may be classified as biological or genetic by an approach centered on measuring environmental correlates of behavioral difference. The consequence has not just to do with classification but with measurement. If "environment" excludes all but the set of environmental variables being measured, and the assumption is that the sources of variance in a population are genetic variation and environmental variation, the approach will generate a different quantitative estimate of genetic variation than will be generated by an approach focused on genetic variation, and vice versa. The results of analysis of variance depend on how the causal space is parsed and how the results of that parsing are measured.[5]

The difference between *effective* and *objective* environment, discussed in chapter 2, provides one example. For quantitative behavioral geneticists, the distinction between shared and nonshared environment is drawn within the set of

---

5. This criticism of analysis of variance is different from the kinds of argument developed by Lewontin (1974) and discussed in chapter 2. I am not rejecting analysis of variance, but pointing out the kinds of assumption required and the method's restricted utility even when the assumptions are warranted.

phenomena they call effective environment—those parts of the environment whose effects can be quantified by measurement of phenotypic similarity and difference. "Shared environment" denotes environmental factors that correlate with nongenetic phenotypic similarities. For social researchers, the matter of interest is what the quantitative behavioral geneticists call objective environment, and that is what they measure directly. In the quantitative behavioral genetics approach, if two siblings whose parents are divorced share a trait, parental divorce is part of shared environment; if they differ, it is part of nonshared environment. In the social-environmental approach, parental divorce is part of the shared environment, whether the siblings are similar or not; nonshared environmental factors are factors that actually differ in the objective environment. So the same factor, parental divorce, can fall into the shared or the nonshared box, depending on which approach is considering it, thus generating incommensurability.[6]

To take another example, in utero effects, if not distinguished from genetic and environmental effects, will fall into the category of environmental effect for a quantitative behavioral geneticist but into the category of genetic (or biological) effect for a social-environmentalist. The concordance estimates that can be derived within each approach will vary as the content of the measured correlate of a given behavioral variation varies. When a possible additional cause/correlate of variation is identified (like in utero factors), it can be separately measured (but of course such phenomena can themselves have either genetic/biological or social-environmental etiologies). But there may be other liminal factors, as yet unidentified, that will affect measurement of identified factors, in the same way unidentified in utero factors affect measurement of identified factors. Analysis of variance can be improved by, for example, explicitly including an interaction factor, but (1) the asymmetries of measurement will persist, and (2) if the interaction is understood as physical rather than statistical, it too will need to be unpacked. Unless there is assurance that the measurement strategies used in different approaches yield the same estimates for the same quantities, the approaches cannot be compared.

Other approaches do not even attempt to measure the correlation of variation in a trait with variation in the range of possible causal factors. They focus instead on understanding the mechanisms or causal pathways more deeply. This interest imposes constraints on experimental design. In particular, researchers' concern with other possible causal factors is focused on holding them constant so as to

6. See discussion in chapter 2 above and in Plaisance 2006, chapter 3. The issues here are analogous to those of the indexicality of representation discussed by van Fraassen 2008.

isolate the operation of the mechanism or causal pathway in question. As the pathways identified by the study of mechanisms narrow, so does the proportion of variation in trait expression explained by a particular pathway. This is dramatically evident in the narrowing of a full-scale integrative approach like DST to the more tractable GxExN approach, which concomitantly narrows the phenomena to be explained from the full range of expression to particular disorders. Finally, as soon as the object of investigation involves interactions, the number of possible hypotheses to be tested (and eliminated) expands exponentially, as do the challenges of designing experiments to sort among them. One may grant that the actual behavioral phenomenon is a function of the intersection of all the causal pathways or systems investigated by these approaches. But each approach offers a partial understanding framed by distinctive questions and distinctive parsings of the causal space.[7] There is no guarantee that approaches will each fit one segment of the range of variation in expression of a trait, that they will carve out exact, discrete proportions of the trait's expression that together sum to one.

We are left, then, with a plurality of approaches generating accounts of the etiology of individual behavioral dispositions that are not reducible to some fundamental level of causation, not integratable into a single comprehensive account, and not empirically commensurable in a way that would permit elimination of rivals in favor of one. This plurality is different from the plurality that results from asking questions posed at different scales, as can be seen by comparing these individualist approaches with population approaches. The causal factors to be investigated belong to different causal spaces to be sure, but the real difference is generated by the phenomena being explained: traits of and variation among individuals, in the one case, and properties of and variation among populations, in the other. A modification of the heuristic grid (fig. 8) illustrates the difference.

Here the approaches focused on individual variation are compressed into one space, reflecting their shared (implicit) commitment to treating distributions of traits in a population, and variation in such distributions among populations, as

---

7. The situation is, therefore, analogous to an example Cartwright (1983, 50) uses to talk about the *ceteris paribus* character of laws. We have a law from one domain of research stating that the higher the altitude, the lower the temperature at which water reaches boiling point, and a law from another domain stating that increasing the saline content of water increases the temperature at which it reaches boiling point. Neither law will by itself explain what happens as we simultaneously move up the mountain and increase the saline content because each belongs to a different explanatory system. In reality, of course, water is always at some altitude and of some degree of salinity. In physics and chemistry we isolate the various causal systems involved and abstract them from the material context they jointly occupy in order to provide explanations.

Genetic Physiology/anatomy	Physical environment	Features of social structure
	(e.g., resource base	(e.g., variations in
Culturally transmitted	shrinking/expanding)	distribution of wealth,
social interaction		of resource access;
patterns		age structure; division
		of labor)

Fig. 8. Causal space for variation among populations. Aggregative approaches in left box; population-level in center and right boxes.

aggregative effects of distributions of individual differences. This distinguishes them from approaches that treat variation in distribution as effects of causal factors, such as physical environment or social structure, operating at the level of populations. Success of the aggregative approach would require that distribution of the causal factors affecting individual traits (genes, school and family environments, neural structures and processes) would have to vary in parallel with the behavioral distributions. Animal ethologists, however, have repeatedly been pushed by their field research to abandon reductionist approaches expressive of such a methodological individualism in favor of population approaches.[8] Consider, for example, a study that sought to understand if there were a sex difference in heritability of aggression, using data from the UK and Sweden. One of the most interesting features of the data the researchers analyze (which they do not attempt to accommodate in their models) is the difference in frequency of aggressive behavior between the two national cohorts. To explain this disparity in terms of the aggregative approach, the Swedish and British populations would have to differ genetically, neurally, or with respect to the patterns of child rearing (or in some combination of these factors) in a way that tracks the different frequencies of aggression in those populations.[9]

Although the moderate point of view would hold that population-level factors interact with those that impinge directly on individuals to produce specific differences, it is just as plausible that in at least some cases population factors swamp individual variation, making the varying contributions of genes, environments, and neural structures to individual differences irrelevant to population differences.[10] A population approach does not deny that individual behaviors are the outcomes of some combination of genetic, environmental, and physi-

---

8. C. Anderson 1990; Barrett 2009; West, King, and White 2003.

9. Eley, Lichtenstein, and Stevenson 1999.

10. See Taylor 2001 for a scenario of this kind.

ological factors or that variation within a population, when physical resources and structural factors are uniform, is ultimately accounted for in such terms. It does reject the suggestion that differences in distributions among populations are the same as differences in distributions within populations.[11] Thus, if we set aside the methodological individualism of the aggregative approach, the plurality evinced by approaches focused on individual traits and on population differences is of a noncompetitive or compatible sort. The two types of approach can be understood as being directed at different scales, and while phenomena proper to the different scales might theoretically interact, there is not the confounding of measurement within the same scale that characterizes the relation of the various approaches directed at individual traits.[12]

Monism or Pluralism?

To demonstrate that a given area of research is presently characterized by plurality leaves open the question of how this fact should be understood from a philosophical point of view. There are several possible philosophical responses to the fact of plurality in a subarea of the sciences. A monist will contend that if a given subarea is characterized by multiple incompatible approaches, this is a temporary phase; there must in the end be one complete and comprehensive account of any given phenomenon or phenomenon type. Forms of moderate pluralism hold either that a plurality of questions supports different and nonreducible, but still compatible, approaches or that pluralism at the theoretical level resolves into an integrated account at the phenomenal level. Both monism and moderate pluralism see plurality as eventually eliminable. Strong or substantial forms of pluralism hold that there are some phenomena and investigative contexts characterized by an ineliminable plurality of theories, models, or hypotheses, and that this situation should not be judged a failure, but should be understood and incorporated into philosophers' understanding of scientific success.[13]

---

11. This point has been made repeatedly in the context of debates over group differences in performance on IQ and other standardized tests, but has not been situated in the kind of pluralist analysis this chapter advocates. See Lewontin, Rose, and Kamin 1984. One point to remember is that "population" does not have a fixed reference: what counts as one population for one purpose may count as several populations for another purpose.

12. It is quite possible that a deeper study of population-level research might reveal the same kind of plurality among approaches focused on population-level factors as I contend characterizes individual trait approaches, but that would have to be demonstrated in a different study.

13. See Kellert, Longino, and Waters 2006, vii–xxix, for extended discussion.

A pragmatist will hold that those approaches that prove valuable in practical (or further cognitive) undertakings ought to be retained. Monism, pluralism, and pragmatism, while signaling philosophical positions, and certainly having both metaphysical and epistemological expressions, are better understood as *attitudes* toward the multiplicity of approaches recounted here. Monism exists primarily as a default assumption underlying polemical and philosophical arguments about these approaches. Pluralism is best understood as an attitude to adopt with respect to the multiplicity of approaches in contemporary sciences. Pragmatism is constituted of a family of specific philosophical positions that have in common a focus on practice and use, but that can nonetheless differ from one another quite markedly.

•

In some cases there is mutual uptake of the sort that suggests integration, as between the developmentalists and the social environment–oriented researchers. But there is no conceptual reason that incompatible approaches, for example, developmental systems theory and quantitative behavioral genetics, could not also, if not collaborate, at least peacefully coexist. This would require recognition of the different kinds of information the two approaches can generate and the different bearing it has on both common and distinctive questions. It would require, in other words, giving up monism. Some analytic concepts, like norm of reaction/range of reaction, may also need to be rethought or more carefully defined, but this does not seem beyond possibility.[14]

In spite of this possibility, in principle, of coexistence, many researchers pursuing one or another of the approaches discussed here adopt a monist perspective in reflecting on their multiplicity. Either they advance their approach as the one best able to account for all of the available data or they expose the weaknesses of alternatives. Advocates of genetic approaches, advocates of environmental approaches, and advocates of developmental systems theory treat the situation as a conflict between intellectual antagonists, the opponent being the more generally, if wrongly, accepted.

Matt McGue, for example, reanalyzes data on divorce.[15] These are interpreted by social-environmental researchers as showing that the experience of being

---

14. This is the theme of Turkheimer and Gottesman 1991. Undoubtedly, the precise claims even these peacemakers wish to make would have to be modified for coexistence to be realized.

15. McGue 1994, as discussed in chapter 3 above.

raised by parents who divorced increases the probability of divorcing in adult-hood. McGue argues to the contrary, that the data support a behavioral genetic interpretation instead. Divorce runs in families because some responsible heritable factor is transmitted across generations.[16] Sandra Scarr treats the social-environmental and DST approaches as equally benighted alternatives to the scientifically informative model of behavioral genetics—the one hobbled by political correctness, the other by misguided determinism.[17] And when Cathy Widom and her colleagues acknowledge that experience of abuse as a child is not 100% predictive of commission of abuse as an adult and leave etiology an open question, DiLalla and Gottesman take her to task for not considering heritable structures (genes) as the causal factor underlying familial reproduction of abuse.[18]

Some researchers engage in extended defenses of their approaches. Steven Pinker is one of the best known among the antagonists. Both a linguistics scholar and a popularizer and defender of behavioral genetics and evolutionary psychology, he has represented the dialectic as a debate between those who would affirm and those who would deny a human nature. His best seller, *The Blank Slate,* is an entertainingly written polemic excoriating thinkers from John Locke to Franz Boas to James McClelland.[19] In the camp of those who would deny a nature are found, apparently, all researchers who investigate the social environment as a source of explanations for individual variation. Non–genetically based behavioral science, he claims, has not advanced since the days of B. F. Skinner:

> Strict behaviorism is pretty much dead in psychology, but many of its attitudes live on. . . . Many neuroscientists *equate* learning with the forming of associations, and look for an associative bond in the physiology of neurons and synapses, ignoring other kinds of computation that might implement learning in the brain. . . . Until recently, psychology ignored the *content* of beliefs and emotions and the possibility that the mind had evolved to treat biologically important categories in different ways.[20]

Even parallel distributed processing (PDP) is reduced to claiming that humans "are just rats with bigger blank slates, plus something called 'cultural devices.'"[21]

---

16. This argument assumes the commensurability, in the sense of common denotation, of "environment."
17. Scarr 1993, discussed in chapters 2 and 6 above.
18. Widom (1989b) and DiLalla and Gottesman (1991), discussed in chapter 3.
19. Pinker 2002.
20. Pinker 2002, 21
21. Pinker 2002, 22

Representing behavioral genetics and evolutionary psychology as embattled truths, Pinker marshals study after study to support the thesis that there is a human nature, that the complexity of human behavior can be attributed to the complexity of the human brain, which is, in turn, a product of the complexity of the human genome. Environmental factors that vary both individually and collectively act on a largely given behavioral repertoire, consisting of some fundamental desiderata and specific dispositions, where desiderata might be goods such as social status, and dispositions tendencies or rules to respond in particular ways to types of environmental features that threaten or enhance the individual's chances of achieving such desiderata. The elements of the repertoire are an outcome of natural selection and transmitted intergenerationally by the genome.[22] To his credit, Pinker incorporates references to many social-environmental studies, but he interprets them from the point of view of evolutionary psychology and behavioral genetics. These interpretations are supported by adaptationist reasoning rather than empirical evidence, however, so the burden of proof remains on the antisocialization arguments. Much of his book, especially when criticizing overgeneralization, sounds quite sensible, but the message of natural selection and genetic determination of basic human psychological and social traits remains its central core. To that message, an obvious response is that the claim that organisms, human or not, should have a biologically based drive to survive is banal, whereas the idea that organisms should have genetically determined plans for specific action in specific kinds of circumstance is a substantial claim crying out for evidence.

A similar argumentative pattern characterizes the opponents of quantitative behavioral genetics. Gilbert Gottlieb demonstrates the inability of genetic explanations alone to account for the establishment of a variety of behavioral patterns in experimental animals as part of his argument for the DST approach.[23] In one instance, he does disavow "environmentalism," but this enters the discussion as a label, not as a distinctive approach.[24] That social-environmental approaches share some of the same shortcomings of behavioral genetics approaches (accounting for only part of the variance, say) gets less attention. Where Pinker's rhetorical stance represents the social-environmental approaches as the hegemonic ones, Gottlieb's treats the genetic approaches as the Goliaths that must be

---

22. The claim is made throughout Pinker's book. For examples, see pp. 51–55 and 101–2 (where the role of genes is specifically championed).

23. Gottlieb 1995.

24. Gottlieb 2001a.

overcome. Diana Baumrind and Eleanor Maccoby, both supporters of hypotheses about the strength of parenting effects on development, likewise represent the behavioral genetic approach as the dominant perspective whose influence must be curbed.[25]

Susan Oyama has done the most to set out a positive theoretical overview of the developmental systems approach, but she too employs a similar kind of binary reasoning, although at a different level.[26] For example, "Change . . . is best thought of not as a result of a dose of form and animation from some causal agent, but rather as a system alteration jointly determined by contemporary influences and by the state of the system, which state represents the synthesis of earlier interactions."[27] The implication of such a formulation is that if change is not one (the preformationist view attributed to genetic approaches), then it must be the other (the systems view), but no consideration is given to the possibility of treating these as two exclusive but nonexhaustive alternatives.[28]

The researchers themselves, then (well, many of them), adopt a dialectical stance characterized by monism: one approach will emerge as the correct one, and it will be comprehensive and exclusive, or if incorporative, fundamental.[29] Results from alternative approaches must be dismissed or shown to be better explained by the author's own preferred framework. Many philosophers have followed the lead of the researchers, either arguing for and against particular approaches or attempting to show how a preferred approach can be made more complete, in order to ward off competitors.

Philosopher Neven Sesardic, for example, intervenes on behalf of behavioral geneticists against environmentalists. He has repeatedly criticized what he takes to be the wholesale condemnation of analysis of variance (ANOVA) initiated by Lewontin and Gould and subsequently repeated by philosophers commenting on behavioral science.[30] He has also taken philosophers to task for constructing a straw man out of Arthur Jensen. The philosophers whom he criticizes have taken Jensen to propose that one can immediately infer heritability (of IQ) between groups from data establishing heritability (of IQ) within groups. Sesardic is careful to point out that Jensen grants that ANOVA must be used with care, but he

---

25. Baumrind 1993; Maccoby 2000.

26. Oyama 1985, 2000.

27. Oyama 1985, 37.

28. These and other mutual criticisms are discussed more fully in Longino 2006.

29. By "fundamental," I mean taken to be principles from which the data and hypotheses of other approaches can and/or should be derived.

30. Sesardic 1993, 2000, 2003; Lewontin 1974; Gould and Lewontin 1979.

is more concerned to defend its utility as an investigative tool and to defend behavioral geneticists' use of it. One of his points is that if, in connection with some trait T, genes are expressed differently in different environments, then the environmentalist is at the same disadvantage as the geneticist in trying to determine the specific etiology of T. Certainly, if ANOVA applied to parent-offspring similarities were the only tool available, this would be the case, but both geneticists (with the advent of molecular geneticists) and environmentalists have more investigative and experimental techniques on which to draw. While he is correct to point out that critics of behavioral genetics have often overstated their case, Sesardic himself is reduced to resting his case for behavioral genetics ultimately on an empirical claim: that gene-environment interactions are very rare. The difficulty for this view is that, even when ANOVA shows high heritability, molecular genetics research has basically shown small, if any, effects for specific genes. In a number of particular cases, as seen in the research of Caspi and Moffitt, discussed in chapter 6, the effects of given genes are greater in those subgroups of study populations exposed to one or another environmental variable, providing data supportive of interaction and undermining Sesardic's claim for its rarity.

By framing the debate in terms of nature versus nurture and as though one of these must be correct, Sesardic is committed both to downplaying the possible contributions of environmentally oriented research and to relying on a highly dubious (at any rate, nonmethodological) empirical claim. For Sesardic, the philosopher's role is to analyze, criticize, and adjudicate. In the adjudicative role, he assumes a monist stance.

A similar stance, but on behalf of other claims, is evinced in the work of Paul Griffiths and Karola Stotz. Both are spokespersons for/defenders of the DST approach, which they present as the alternative to excessively gene-centric approaches in biological and behavioral research. Griffiths is particularly interested in using developmental research to open up conceptions of evolution, hoping to move it away from the neo-Darwinian consensus that treats genes as the units of heredity to a more expansive conception. Stotz is more focused on genetic research and concerned to stamp out the remnants of the Central Dogma of DNA first enunciated by Francis Crick.[31] She argues that the role of epigenetic factors and processes in gene activation and regulation shows that genes play no causally privileged role, that all factors have equivalent roles to play in the

---

31. There are varying formulations of the Dogma. Stotz addresses the idea that information (and causation) proceeds in only one direction, from genes to RNA to proteins to higher-level traits.

causal processes of development. Even the specificity of linear amino acid se-quences is a product of multiple actors within the cell.[32] In a 2008 article, her target is the view that "the gene is deterministic in gene expression and therefore all gene products are fully specified by the DNA code."[33] The principal object of her criticism is C. K. Waters's account of genetic causality, which utilizes James Woodward's account of causation as difference making. Waters uses a concept of "actual difference making" to put the relation of a segment of DNA involved in a synthesis event to its immediate product (a linear segment of RNA) in a philosophical context.[34] It is the sequence specificity of the DNA that is respon-sible for the specific sequence of bases making up the RNA. Other molecules are required to effect the synthesis, including the so-called splicing agents that facili-tate selection and combination of DNA segments, but Waters emphasizes that it is the linear sequence of bases in the DNA segment(s) that is "the actual differ-ence maker," determining that the RNA segment consists of sequence s_1 rather than s_2. This role explains the intense interest of molecular biologists in DNA. (It also explains why the metaphors of "code" and "information" have gained such purchase both inside and outside the molecular biological community.)

Stotz worries that granting even this much supports a reductionist concep-tion of the organism and of development. Even as she seeks to rise above the nature-nurture dichotomy/debate, she recreates a reductionist/DST dichotomy. Unless one is prepared to accept the multilevel, omnifactorial approach of DST, one is committed not only to reductionism, but to "preformationism." To accuse geneticists and philosophers of preformationism in this context is to construct a straw man and to be in the grip of a monist conviction that only one approach can be correct. In addition to collapsing the alternatives to create another prob-lematic dichotomy, Stotz is collapsing metaphysics and empirical investigation. Without denying that the role of DNA has been vastly overblown, especially in public statements by and about molecular researchers, it should be possible to see that the linear specificity of DNA gives it a role of particular interest in the organism. And had that role not been the subject of intense investigation at the molecular level, the complexities of gene regulation and RNA and protein synthesis would not have been discovered. DST offers a picture of the complex-

---

32. Stotz seems not to make a distinction between there being some specific sequence and the sequence being the sequence it is as opposed to being some other sequence. Splicing does "select" parts of the DNA molecule for amino acid assembly, but it is the sequence of bases in the spliced molecule that accounts for the amino acid sequence.

33. Stotz 2008.

34. Waters 2007.

ity of the organism that is probably correct from a metaphysical point of view and that serves as a prophylactic against the equally metaphysical temptations of reductionism. But the biological processes that constitute this complexity are discovered by contrastive experimentation, one molecule/interaction at a time.

Monism, then, is supported neither by the empirical results to which research can point, nor by the arguments of philosophers who presuppose it. In fact a monist attitude leads them to misinterpret the phenomena about which they write. From an empirical point of view, what we know is piecemeal and plural. Each approach offers partial knowledge of behavioral processes gleaned by application of its investigative tools. In applying these tools, the overall domain is parsed so that effects and their potential causes are represented in incommensurable ways. We can (and do) know a great deal, but what we know is not expressible in one single theoretical framework.

This epistemological situation has stimulated some practitioners to consider alternatives to single comprehensive models. Even as academic researchers argue about the relative adequacy of their competing views, clinician-researchers must contend with the multiplicity of frameworks that might guide their practice. Psychiatrist and behavioral geneticist Kenneth Kendler, whose work was discussed in chapter 6, considers the problem of diagnosis and treatment in the context of the multiplicity of causal factors and the apparent incompatibility of models representing multiple causal systems. In one article, he demonstrates the nonaggregative character of the interactions among genetic, physiological, and psychological-environmental factors to argue against reductionist causal stories in favor of what he calls a mechanistic approach.[35] In Kendler's approach, multiple mechanisms are involved in the production of behavior. Research focused on single mechanisms is required in order to understand each type of system. But the knowledge thereby achieved is partial. More comprehensive understanding relevant to and required for treatment will be accomplished only through integration: understanding how these mechanisms interact in the production of specific diseased or disordered states. The form of these interactions will vary from condition to condition, perhaps from individual to individual, so no single model encompassing all behavioral pathologies, let alone behavior in general, can be expected.

Philosophers too are investigating pluralism as a way to address the plurality of models. Philosophers' pluralism comes in different forms, however, from

35. Kendler 2008.

weak to moderate to strong. Both James Tabery and Sandra Mitchell advocate moderate forms of pluralism. Tabery, in a review of Sesardic's *Making Sense of Heritability*, diagnoses the conflict between quantitative behavioral geneticists and their antagonists as a matter of belonging to different research traditions asking different questions and thereby committed to different understandings of GxE interaction.[36] The quantitative behavioral geneticists belong to a biometric tradition extending back to R. A. Fisher and concerned with understanding variation in a population, while the developmental systems theorists belong to the developmental tradition extending back to Waddington and concerned with individual development. Tabery considers a proposal that we understand the different traditions as operating at different levels of analysis, such that their concepts of GxE interaction will of necessity be different. It is a statistical concept for the biometricians, referring to what remains to be explained when variation in a trait in a population has been apportioned to biological and environmental variation taken independently. It is a causal concept for the developmentalists, expressing the interaction of heritable (genetic) and environmental factors within the organism as it develops through its life stages. Tabery points out that this solution presupposes that there is no leakage from one level to another. He argues that the causal GxE interaction identified by the developmentalists is a factor in population-level variation and that GxE as a statistical remainder within population variation represents the interdependence of difference makers.

Tabery's solution suggests that the population and developmental approaches are reconcilable, that each identifies factors relevant to the other, but from a different vantage point. If this solution means, however, that the GxE interaction is effective only for the proportion of variance in a population not accounted for by G and E taken separately, it will not satisfy the developmentalists who argue that their form of interaction characterizes every individual. Nor will it satisfy the quantitative behavioral geneticist who argues that the causal processes internal to individuals do not illuminate patterns of variation within a population. Even more seriously, Tabery's proposal seems to assume that measurement is not affected by the vantage point from which the phenomena are measured. But this is just what the previous section shows to be the case. To the extent that Tabery's is a form of pluralism, it is a moderate pluralism too weak to accommodate the incongruence of causal spaces investigated by the different approaches. It is appropriate to the plurality of explananda that characterizes the difference

---

36. Tabery 2009.

between investigations of intra- and interpopulation variation, but not to the plurality that characterizes the difference among approaches focused on individual difference.

Sandra Mitchell advances a position she calls "integrative pluralism," which contrasts two forms of pluralism: competitive and compatible.[37] Competitive pluralism is the maintenance or persistence of multiple mutually exclusive accounts of the same phenomena. It has been described and defended as a strategy a research community can adopt to hedge its bets in the face of empirical uncertainty and as a strategy that facilitates the exposure of empirical weaknesses. Both defenses see plurality as a stage of inquiry aimed at identifying the one true or best theory. Competitive pluralism is thus, at heart, a monist strategy. Compatible pluralism is Mitchell's name for a noncompetitive form of plurality. One of its forms understands a plurality of models or theories of a single phenomenon as a consequence of questions posed at different levels of analysis, organization, or granularity. There is no competition, but neither is there any troubling persistence of mutually exclusive alternatives.

Mitchell argues first, like Tabery, that these levels are not isolated from one another and that processes at one level can constrain the possibilities at other levels. Secondly, she argues that plurality within a level need not be competitive, that is, need not involve mutually exclusive alternatives. Her argument for the first is the interdependence of factors accounting for the division of labor among social insects (ants, bees). Adaptationist explanations ignore developmental constraints and treat the observed pattern of division of labor as the outcome of natural selection's eliminating inferior variants. But developmental constraints may limit the kind of variation that would have been available for natural selection to work on. As for plurality within a level, she argues, focusing on several different self-organization models of division of labor, that each model is an idealization that applies to a system in which only the modeled factors are at work. In actual systems, multiple factors or processes may be at work, in which case the system must be understood as one in which the factors/processes are integrated.

At this point, however, integrative pluralism faces difficulties similar to those that beset developmental systems theory. While it may become possible to model how such integration would work, as soon as more than a few processes are involved it becomes empirically indeterminable which of many possible integrations is actualized in a given system. We may remain metaphysically commit-

---

37. Mitchell 2002, 2009.

ted to the view that multiple factors or processes are interacting in our one world without being able to demonstrate the relative superiority of any one model of their interactions. And integration may require omitting factors and thus losing knowledge of their role. Epistemologically, we may learn more about a system by utilizing multiple partial representations, each of which enables us to go further in our study than would the attempt to obtain a complete representation of all causal interactions. Our understanding of the system may require not the integration of the different models but acknowledgement that each represents one aspect of the system. Mitchell's pluralism is guided by monist intuitions: one system-one complete model integrating multiple processes. But this presupposes a commensurability that may not obtain in all cases, and does not obtain in the case of behavior.

I would distinguish the forms of pluralism differently than do Tabery or Mitchell. Instead of competitive and compatible, I would think in terms of eliminable or ineliminable. Eliminable plurality is plurality that is temporary, a way station to a single comprehensive account. Eliminative pluralism is the view that all plurality is of the eliminable sort. Ineliminable plurality is plurality that results either from the availability of models at different levels of analysis or organization or from the incommensurability of equally empirically adequate approaches at the same level of analysis. Non-eliminative pluralism is the view that some plurality is of the ineliminable sort and, importantly, of the second, incommensurable-approaches sort. Non-eliminative pluralism is an attitude toward that plurality of incommensurables, an attitude focusing on the different kinds of knowledge each approach can offer rather than assuming that one, at most, is correct. It has the advantage of allowing us to see what kind of knowledge can be generated by what kind of methods. It faces several challenges, however. One is the task of explaining the sense in which each of these approaches is correct. What can it mean to say that two or more approaches are correct if they assign different and incompatible values to the same parameter? A second problem concerns the question of how to design scientifically informed intervention and policy in the face of multiple, and *prima facie* incompatible, research approaches emphasizing different causal factors or combinations.

In *The Fate of Knowledge*, I proposed an umbrella concept for the various kinds of epistemic or semantic success attributed to scientific representations.[38] Conformation (of a representation to its object) encompasses truth (literal) at one extreme, but also isomorphism, homomorphism, similarity, approximation,

---

38. Longino 2001b, 115–21, 136–40.

and other relations that name forms of success in representation. Unlike the categorical dichotomy true/false, these other forms of success admit of degree, and their application requires a specification of respect(s) in which they hold. To say that two different representations of the same phenomena are equally correct is to say that for each there is a degree and respect in relation to which it can be said to conform to its subject matter.

Because all of the approaches reviewed produce representations that at some level conform to some degree and in some respect to their common subject matter, conformation may seem too coarse a term of success. But the bar for success is set by specification of degrees of and respects in which conformation is sought, not conformation pure and simple. In comparing claims issuing from different approaches, it is necessary to use at least a common degree of conformation. But if the degree is determined in different measurement setups—that is, different parsings of the causal space—conformation to degree n in one setup is not equivalent to conformation to degree n in a different one. Take the problem considered earlier for both genetic and environmental explanations: that each not only ignores in utero, gestational factors, but to the extent that they apportion variance between genetic and environmental factors, each must put uterine effects in what is for it the alternative category. The values assignable to G or to E will vary depending on what is included in those categories. This means that correctness must be relativized to the initial parsing of the causal space. Hypotheses articulated in relation to a common parsing of the space can be compared empirically, but it is not possible to compare and adjudicate empirically among hypotheses articulated in relation to different parsings of the potential space. Understanding the phenomena involves understanding the scope and limits of and relations among the range of ways we have of developing knowledge of them.

It is, of course, possible to engage in meta-empirical evaluations: How much of the explananda can be accounted for? How parsimonious are the explanatory frameworks? These meta-empirical criteria are encompassed in the respects in relation to which representations may be said to conform. They are meta-empirical insofar as setting any particular benchmark of scope or simplicity is not itself an empirical matter. They can be applied to bodies of data and their explanatory models generated within a particular approach, but they are not themselves the outcome of or defensible by appeal to observation or experiment. Among respects relative to which conformation is to be evaluated must be included some specification of what the results of investigation are to be used for—whether to feed into further empirical research, to support a big picture, or to guide strategies for intervention in or prevention or enhancement of behaviors or conditions. Different kinds of conformation will be mandated by different pragmatic aims.

This application of the notion of conformation to the behavioral research considered here means that the second problem for pluralism, the problem of designing policy or intervention in the face of multiple approaches (i.e., the problem of application), arises in the context of a flawed model of scientific knowledge—a model that separates pure knowledge from its application and supposes that "pure" (a.k.a. "basic") research can provide comprehensive knowledge of a phenomenon that can then be applied to or drawn on for the solution to practical problems. To the contrary, the practical problems and their associated constraints shape the criteria involved in the evaluation of research results. Research cannot be separated from conceptions of what we want the resulting knowledge for. This does not collapse into a crass pragmatism that equates truth with utility. The issue is that there are many truths, and the task is to sort out which of the truths is useful to which purpose.

If this is the task, pluralism must be supplemented by a form of pragmatism. Another of the Kendler group's papers exemplifies the kind of pragmatically inflected pluralism I am advocating.[39] Here the authors draw on James Woodward's account of causation as difference making (which they term "interventionist") to argue that understanding particular psychiatric disorders is best achieved by framing studies around interventions that increase or reduce the incidence of a particular disorder. This interventionist model allows for the identification of difference makers within multiple causal systems and stays focused on the practical "goals of psychiatry which are to intervene in the world to prevent and cure psychiatric disease." The pragmatism of Kendler's interventionist and mechanistic models is intended to map guides to action that do not privilege any particular causal mechanism. He faults a vision of scientific understanding that places a premium on the discovery of laws for engendering sterile debates between approaches. One could also understand his rejection of reductionism as a repudiation of monism. In his model, the conditions themselves, and interventions empirically demonstrated to make a difference to their incidence, drive the development of understanding. The plurality of types of causal factor and of causal mechanisms remains an inescapable feature of the domain of scientific psychiatry.

The type of success envisioned by Kendler is limited to the identification of and ameliorative intervention on factors in psychiatric disorders. The research

---

39. Kendler and Campbell 2009.

approach is not intended to lead ultimately to a comprehensive theory of human behavior. The implicit picture is instead that of a system—the person—whose manifest behavior is a function of complexly interacting factors. Disorders or malfunctions of the system may have singular causes or be the outcome of the intersection of several causal subsystems. But knowledge of what produces a malfunction is not equivalent to knowledge of what is involved in successful functioning, nor even to knowledge of how the factor responsible for malfunction otherwise contributes to successful functioning.[40] The lesson from Kendler is again the lesson of partial knowledge: knowledge of how isolable causal subsystems work and knowledge of how different subsets interact in different instances of malfunction. Partial knowledge is no less knowledge for being partial.

The political critics of the various approaches have misdirected their ire. The problem is not the effort of any particular approach to refine and extend the understanding its investigative methods make possible. The problem is the supposition that general laws of human behavior might be discoverable, that one of the causal subsystems might be dominant, or that one investigative approach might offer the fundamental explanatory model or theory. To say this is not to exonerate the behavioral sciences completely of problematic social consequences. The problems lie not in the efforts to identify causal factors but, if the analysis of the following chapter is correct, in the conceptualization of the behaviors themselves.

---

40. I set aside the question of determining what counts as a malfunction and what a mismatch of behavior and expectation.

9

Defining Behavior

P revious chapters have focused primarily on differences among approaches to the etiology of behaviors and behavioral dispositions and the methods for studying them. In this chapter I take up a second major methodological question: the characterization of behavior for purposes of scientific study. A recent study by behavioral biologist Daniel Levitis and colleagues reviewed a variety of kinds of publications (from textbooks to research articles), identified a series of key points in the conceptualization of behavior, and then canvassed colleagues in a variety of behavioral biology subfields.[1] The authors found no unanimity in the definition of "behavior" across texts or individual researchers. A similar disunity characterizes the conceptualization of individual behaviors, at least those on which I have concentrated here. There are at least three distinct issues: (1) how the specific behaviors under study are operationalized, (2) how behavior generally is understood, and (3) what the units of analysis are.

The Phenomena

Research that takes a specific category of behavior as its explanandum starts with concepts that belong to our moral economy—folk psychological concepts. How could it not? Nevertheless, our ordinary concepts are not well suited to

---

1. Levitis, Lidicker, and Freund 2009.

scientific investigation, being vague and inflected with the social values that make the behaviors they designate salient in the first place. One of the challenges for researchers, then, is the development of clear and unambiguous behavioral categories and criteria. Attempts to provide these still bear traces of the social context that researchers hope to expunge. Recent research on aggression in particular has made links to concerns with crime, antisocial behavior, and violence. This marks a change from the 1960s and 1970s, when much of the research on aggression was related to research on sex differences in behavior, and often used to support claims that male dominance in human and nonhuman societies was natural. While some anthropologists treated aggressiveness as a possibly obsolete carryover from early hominid biology, for other thinkers it was associated with other positively regarded, putatively masculine characteristics, such as leadership. So much has the valence of aggression research changed that in some instances aggression is operationalized as violent criminality, as evidenced by conviction rates. While operationalizations for research purposes seek to make the concept precise, this remains a very broad notion, covering many different behaviors.

What seems to warrant classifying a behavior as aggression is the direct, personal infliction of harm. Only some forms of such behavior warrant scientific study, however. Many forms of harm, whether directed or the by-products of other actions, are not studied. Research on aggression, and more narrowly, criminal violence, seems designed to address only general public concerns about unsanctioned violence intentionally perpetrated by and directed toward individuals—the prototypical violent crime as dramatized in local news reporting. As noted in chapter 1, the concern of some critics of aggression research, especially biological research, is that in a society already characterized by racial antagonism, such research will be used to forge a link between racial identity and aggression, just as similar research was employed to forge a link between racial identity and intelligence.[2]

Research on sexual orientation also concentrates on a dramatic difference, in this case, same-sex/gender versus other-sex/gender erotic focus. As remarked in chapter 1, "sexual orientation," in this research context, is really code for homosexuality. Nevertheless, the research, broadly understood, serves multiple functions, aimed at understanding a range of reproductive behaviors, sexual function

---

2. Of course, these putative links are much disputed (Gould 1981; Lewontin, Rose, and Kamin 1984; Lewontin 1991). Once established in the public imagination, however, they are difficult to break, as the research of Eberhardt and colleagues demonstrates (Goff et al. 2008).

and development, patterns of intimacy, and, for some, social deviance. In work on nonhuman animals, it is part of understanding the range of reproductive behaviors, including mate selection, courtship, intercourse, and parenting. In humans, it tends to be part of programs studying intersexuality and hormonal dysfunction or deviance. Until the rise of the human sciences in the nineteenth century, sexual behavior, including homosexuality, was a moral, religious, or legal matter. During the nineteenth century, homosexuality came to be seen as a psychiatric disorder. While no longer conceptualized as a disorder (except, perhaps, by a religious fringe), homosexual behavior is still intensely studied, with respect both to its frequency and to its etiology. Sex researchers, like other behavioral researchers, are keen to shed the judgmentalism of the past, but their efforts have had mixed success outside of the research context. In what might be deemed a return to the premodern, homosexual acts are condemned as sins, and gay men and lesbians are still targets of hostility and unprovoked assault, as well as of efforts to get them to repent and "reform."

Research on aggression concerns behavior that is in principle and in reality engaged in by persons in all subcategories of society (racial, sexual, economic). The political concern about biologically oriented aggression research is that it will be used in the service of sociopolitical agendas that deliberately or coincidentally target one segment of society, that is, that a category of *aggressive* or *criminal* emerging from the research will coincide with another social category, for example, poor black and Latino men. By contrast, homosexual behavior is already associated with a social category—homosexuals—so there is no question of pinning it on a scapegoat.[3] For gay men who have welcomed the research, it is seen as putting sexual and erotic orientation beyond the domain of choice, and hence outside the domain of will or sin. If sexual orientation is determined, it is not an object of choice or decision, and it makes no sense to worry about the recruitment of young people to "the gay lifestyle," as some fear. Others, however, assess the work in the context of the general therapeutic aims of biomedical research. In their view, biological knowledge could as easily be used to support fetal deselection or other (pre- or postnatal) biological interventions to reduce or eliminate homoerotic urges as to argue for tolerance.

---

3. Historical and cross-cultural studies show that homosexual behavior has not always and everywhere been associated with a special of category of persons, i.e., that this identification is a culturally specific one. Michel Foucault and gay historians have discussed the emergence of the category "homosexual" in the late nineteenth century. One might say that some of the resistance to the biological study of aggression is an attempt to prevent the emergence of a similar category of "aggressive," and especially to prevent the emergence of such a category that is racially inflected.

The analysis in the preceding chapters of various approaches to the study of behavior should undermine any worry about the imminent effectiveness of the work in biosocial engineering focused on racial, ethnic, or gay populations. But study of the eugenic and therapeutic programs of the last century shows that concepts associated with research can have a life independent of the empirical confirmation or disconfirmation of hypotheses employing those concepts. Both aggression and sexual behavior have undergone changes in valence in the last fifty years. What has not changed is the fact of their social charge. Whether that charge is positive or negative, they remain topics of intense concern. This concern pervades the environment of their scientific study. In this chapter, I look more closely at the characterization of these behaviors as objects of explanation, and integrate concerns about those characterizations with recent philosophical reflections on the general concept of behavior.

Scientific Definitions of Aggression and Sexual Orientation

One critical methodological issue in behavioral research is the identification and definition of behaviors, not just of the causal factors contributing to their expression. A second is how to measure and establish the relatedness of the phenomena under investigation. Aggression and sexual orientation differ in that the study of aggression is really the study of a host of related behaviors, whereas the almost universal adoption of the Kinsey scale makes sexual orientation seem more straightforward. Things are, however, more complicated in both areas.

Aggression

The phenomenon of aggression has been well studied in a variety of animal models. Research programs in behavioral endocrinology have investigated the role of organizing and activational hormone secretions in enhancing or diminishing a variety of well-defined behaviors classified as forms of aggression. These include stereotyped motor sequences, such as flank marking, as well as responses to intruders in a cage and maternal defense of young.[4] Human aggressive behaviors are more difficult to study. They do not include easily identifiable stereotyped sequences, nor is it any longer permissible to set up experimental situations in which violence or aggression might be elicited.[5] Thus, in human studies,

---

4. For a recent review, see Simon and Lu 2006.
5. Milgram 1974.

aggression or aggressivity (i.e., an aggressive disposition) is either identified ret-rospectively, through various means of counting past instances, or through sur-rogate behaviors treated as diagnostic of aggression. In their review of studies of testosterone and aggression, Albert and colleagues argue that human aggression falls into the category of defensive aggression, which is a response to a perceived threat.[6] It is thus cognitively mediated, further complicating its study. I use the idea of operationalization to refer to the specific kinds of phenomena that are measured as indices of aggression or aggressivity. Although certain operational-izations occur more frequently in certain approaches, there is no consensus on a neat partition. Researchers working in each of the approaches employ different indices and operationalizations of aggression and different methods of observa-tion. Treatment of any of these as having a bearing on aggression assumes signifi-cant overlap among them.[7]

Sexual Orientation

Sexual behavior, too, has been well studied in animal models. The hormonal system governing the expression of lordosis in rats (concavely arched back, pre-sentation of rump) was exquisitely delineated by Donald Pfaff.[8] The hormonal systems involved in the expression of mounting behavior have been studied in various species. Lordosis is female-typical, while mounting is male-typical. The ability to increase or decrease the expression of these behaviors through the manipulation of hormones at critical developmental periods has led to the pro-duction of animals of one chromosomal sex that express behavior typical of the other sex. Such cross-sex behaviors have been proposed as a model for human homosexual orientation. There are several difficulties with this suggestion, apart from the fact that it treats human sexual behavior from a perspective concerned primarily with procreation.[9] One is that the intended human analog is not (or not primarily) a matter of specific sexual acts or stereotyped motor sequences, but rather a matter of perceived features of the sexual partner with whom those acts are performed. A second difficulty is that human sexual/erotic orientation

---

6. Albert, Walsh, and Jonik 1993.

7. One significant line of research concerns establishing the validity of various measures. See Buss and Perry 1992; Orpinas and Frankowski 2001; Halperin, McKay, and Newcorn 2002; Weinshenker and Siegel 2002. Nevertheless, researchers still classify subtypes quite differently. See Eley, Lichtenstein, and Stevenson 1999 for some examples.

8. Pfaff 1980.

9. This particular criticism is developed by Gooren 1990.

is at least as much about intimate partnering and love as it is about sex acts or sexual gratification. The recent work on the induction of male-male chains in drosophila partially addresses the first of these issues but not the second.[10] Furthermore, while many researchers on human sexuality continue to employ the Kinsey scale, others find it phenomenologically inadequate. As I will make clear later in this chapter, human sexual behavior is characterized by many dimensions of variation, raising the question of why one rather than another should be chosen for investigation.

What Is Aggression?

Two categories of literature, research reports and research reviews, give different kinds of answer to this question. Research reports present a single study using a single research methodology on a particular population. Research reviews and meta-analyses either review the results obtained using one approach in a variety of studies or attempt a comparative assessment of studies using different perspectives. Articles reporting on cross-indexical concordance constitute a third category, either supporting the use of given indices and measurement strategies or calling them into question when concordance fails.

Research Reports

As we have seen, studies conducted on particular populations or on particular causal factors attempt to establish a relation, ideally a causal relation but minimally a correlation between some state—genetic, social/familial, physiological—and some form of the behavior of interest. Two problems confront the reader perusing a selection of this work. One is the variety of behaviors that count as aggression. The second is the variety of measurement strategies employed.

•

What is counted. Laboratory research on animal populations studies behaviors like flank marking, offensive aggression against intruders to a cage, maternal aggression, and sexual competition. Many of these studies manipulate hormone levels, especially gonadal hormone levels, associating variation in frequency of

---

10. See Zhang and Odenwald 1995. It should be noted, however, that this work is about a very restricted component of the range of same-sex sexual interaction.

one of these behaviors with variation in doses/levels of a given hormone administered perinatally (as an organizing factor) or contemporaneously (as an activating factor). The animal behaviors observed are the behaviors about which information is sought, though one must remember that laboratory animals are bred to be more physiologically uniform than their wild counterparts. The situation is quite different for human behavior, where researchers must make do with surrogates for the behaviors or dispositions they seek to understand. These include

- conviction of violent crime;[11]
- fighting;[12]
- delinquency (in minors);[13]
- violent rage;[14]
- anger, irritability, and verbal aggression;[15]
- hitting a doll (in a psychological observation setting);[16]
- scores on psychological measuring instruments like the B-D Hostility Inventory or the Aggression subscale of the Child Behavior Checklist or the Grey and Cloninger checklist;[17] and
- diagnoses of antisocial personality disorder or oppositional defiant disorder as per the *Diagnostic and Statistical Manual of Psychiatric Disorders* (*DSM*, now in its fourth edition).[18]

Looking at these measuring and diagnostic instruments further fragments the concept. In the *DSM III* (revised), antisocial personality disorder (APD) is introduced as follows: "The essential feature of this disorder is a pattern of irresponsible and antisocial behavior beginning in childhood or early adolescence and continuing into adulthood. For the diagnosis to be given, the person must be at least 18 years of age and have a history of Conduct Disorder before the age of 15."[19] In addition, the individual must exhibit at least *four* of the following:

11. Raine, Brennan, and Mednick 1994; Widom 1991.

12. Haapasalo and Tremblay 1994. This and the citations in the following six notes represent only a small sample of the relevant literature.

13. Widom 1989b.

14. Heiligenstein et al. 1992.

15. Heiligenstein et al. 1992; Coccaro, Gabriel, and Siever 1990; Bierman and Smoot 1991; Lavigueur, Tremblay, and Saucier 1995.

16. Plomin, Foch, and Rowe 1981.

17. Bierman, Smoot, and Aumiller 1993; Luntz and Widom 1994.

18. Coccaro, Silverman, et al. 1994; Speltz et al. 1995.

19. American Psychiatric Association 1987, 342

- inability to sustain consistent work behavior, indicated by (1) significant unemployment for six months or more when work is available, (2) repeated absences from work unexplained by illness, or (3) abandonment of jobs without realistic plans for others;
- failure to conform to social norms with respect to lawful behavior (repeatedly performing antisocial acts that are grounds for arrest);
- irritability or aggressiveness, indicated by repeated physical fights (includes spouse- and child-beating);
- repeated failure to honor financial obligations;
- failing to plan ahead, impulsivity as indicated by (1) traveling from place to place without a clear goal for the period of travel or (2) lack of a fixed address for a month or more;
- disregard for the truth (lying, use of aliases, "conning" others);
- recklessness regarding one's own or others' personal safety (driving too fast or while drunk);
- failure to function as a responsible parent;
- failure to sustain a monogamous relationship for more than one year; and
- lack of remorse.

The diagnostic criteria in the more recent *DSM IV* reduce the number of behaviors required for a diagnosis of APD to three and eliminate sexual promiscuity as a criterion.[20] But clearly, an individual can be correctly diagnosed with APD without ever engaging in physical violence. Turning to the items in the list of behavioral phenomena that are the objects of studies, it is not evident that they are expressive of some single underlying disposition. They might as easily be a heterogeneous collection of behaviors that we don't like to be the recipients of or that are incompatible with the expectations of workers in industrial and postindustrial society. Some argument is needed that they constitute a coher-

---

20. American Psychiatric Association 1994. Both the *DSM IIIr* and the *DSM IV* make a point of distinguishing antisocial personality disorder, one of the targets of study in both environmental and physiological (serotonin-uptake) research, from adult antisocial behavior, which is criminal or aggressive behavior and does not meet the full set of criteria for APD. One wonders, then, what relevance the research will have for criminology. The *DSM V*, currently in preparation, introduces significant changes in the criteria, but only the diagnostic criteria of *DSM III, IIIr*, and *IV* would have been in use in the research reported on here. See *DSM* revised criteria, http://www.dsm5.org/ProposedRevision/Pages/proposedrevision.aspx?rid=16 (accessed June 15, 2011).

ent phenomenon.[21] Furthermore, the role of *DSM* categories in establishing the legitimacy of treatment for insurance purposes should raise some concern about their aptness for use in research contexts.

•

Measurement strategies. The second challenge to assessing this work is the variety of measurement and observational strategies. For animal models, not only are the behaviors of interest directly observed, but there are standardized procedures or protocols of observation. Also, with few exceptions, the behaviors under study are stereotyped and easily identifiable and countable, for example, biting or nipping at the flank of another male. The sample human populations, which include both children and adults, are either ready-made in clinic or prison settings or constituted of volunteers who respond to calls for subjects in psychological studies. The propensity to aggression operationalized in one or more of the ways noted above is determined by

- arrest and conviction for assault or other violent behavior;[22]
- self-report;
- other report (parents, siblings, peers, teachers);
- answers in structured interviews;
- responses to psychological inventories (such as the MMPI) or to standardized narrative or visual stimuli; or
- psychiatric or other clinical diagnosis.

Finally, the concept of environment is not stable across these studies, whether they are sociologically oriented (and attempt to show environmental influence) or biologically oriented (and attempt to show the role of biological factors as against environmental factors). The environment may consist of

---

21. Of course, efforts to establish cross-measure uniformity/reliability (see note 7) count as one step in such an argument, but only one. These tend to focus on establishing validity across particular survey instruments, rather than across the variety of operationalizations of aggression.

22. Given the widespread acknowledgment of the role social factors like race and socioeconomic class play in arrest and conviction, it is surprising that the latter continue to be treated as criteria for aggressivity. Without a means of segregating the role of such biasing factors in the construction of the sample population, studies relying on arrest and conviction can't be regarded as informative about aggressivity generally. They are informative only about susceptibility to arrest or conviction.

- general features of the rearing environment (including home, school, media exposure, peers);
- particular features of the parent-child interaction;
- everything that correlates with behavioral difference or sameness that's not genetic—the uterine environment, physiological environment within the organism, and social-familial environment, and the "effective environment," shared or nonshared, of the behavioral geneticists; and
- socioeconomic status and surroundings.

Clearly the studies themselves constitute a kaleidoscopic patchwork. Each attempts to establish a correlation between a purported causal factor and some behavioral or dispositional phenomenon that bears some resemblance to aggression or is commonly taken as a marker of aggressive behavior. Not only do the measurable target phenomena differ, but, given the variety in conceptions of "environment," the ontologies of the domains are differently structured. Unless, however, competing genetic and environmental accounts, for example, are accounts of the identical phenomenon, it's not clear what it means for them to constitute competitive accounts. As was remarked of the causal ontologies identified in the previous chapter, unless studies target the same behavior, assessed in the same way, using the same measuring instruments and the same concepts of gene, individual, and environment, it's not clear that any can be taken to contradict or complement or replicate any other.[23] The second category of literature, research reviews and meta-analyses, is intended to address this problem. These attempt to differentiate and classify the behaviors studied, with a view to articulating what general conclusions they might collectively support.

Research Reviews

The review by Albert, Walsh, and Jonik on biological foundations of aggression in humans is mainly devoted to analyzing studies of the relation of testosterone levels to aggressive behavior, and articulating just what the implication of

---

23. This, of course, is one of the points of the Levitis, Lidicker, and Freund (2009) critique of the multiplicity of conceptions of behavior in behavioral research. Turkheimer and colleagues have documented the mismatch between self- and informant behavioral assessment (Klonsky, Oltmann, and Turkheimer 2002; Clifton, Turkheimer, and Oltmann 2007). Finally, a meta-analysis of aggression studies reported finding strong heritability using self- and parent reports to identify aggressive individuals, but a strong environmental effect when using laboratory observation to identify aggression (Miles and Carey 1997). Both kinds of mismatch cast doubt on the coherence and robustness of the construct.

animal studies is for human behavior.[24] This paper distinguishes between types of behavior on the basis of the neurological control systems involved. This results in a tripartite classification:

- hormonally modulated aggression (males against male conspecifics or, in the case of females, maternal aggression against threats to offspring);
- predatory aggression (primarily against animals of other species constituting the predator's diet); and
- defensive aggression (a response to a perceived threat).

Because both hormonally mediated aggression and predatory aggression are highly stereotyped motor behaviors or behavior sequences particular to a given species, Albert's group argues that they are not an appropriate model for human aggression, which does not follow standard, stereotypical patterns but varies in expression both within and among individuals. They conclude that defensive aggression provides the best model for human aggression. This form of aggression can be expressed at any age, requires a specific activating experience, is directed both at familiar and unfamiliar individuals and at both males and females, and shows no inherent gender differences (i.e., any sex difference in expression of defensive aggression is a function of the differences in situation). Defensive aggression is cognitively mediated, that is, it involves some kind of judgment. The best hope, then, according to Albert and colleagues, for identifying biological foundations for defensive aggression lies in studying the involvement of brain areas.

This attempt to bring order to human aggression studies through the lens of animal studies does not directly address the variety of criteria used to identify aggression. By focusing on the classifications made possible by animal studies, all human aggression gets typed as of one sort. While the review suggests the irrelevance of much work with animal models to the human case and points to the kinds of biological study that would be relevant, it does not help to assess the existing range of human studies.

The review by Mason and Frick of work on the heritability of aggression does focus specifically on at least some of those human studies, and, in particular, attends to the significant variety of behaviors classified as aggression for purposes of conducting scientific studies. Indeed, one of their aims is to "test whether the estimated genetic effects are influenced by the definition of antisocial behavior

---

24. Albert, Walsh, and Jonik 1993.

that is employed."[25] They proceeded by conducting a meta-analysis of twin and adoption studies designed to determine the heritability quotient of aggressivity. As noted in chapter 2, Mason and Frick's literature scan identified seventy such studies published between 1975 and 1991. In order to guarantee the quality of the data that went into the meta-analysis, they developed a set of selection criteria that excluded adoption studies that did not compare concordance rates of adoptees and biological parents with those rates for adoptees and adoptive parents; studies that confounded antisocial behavior with "other forms of psychological disturbance"; studies that did not compare subjects on the same type of behavior; and studies based on subsamples from studies already included. Applying these criteria to the initial set of seventy studies left them with fifteen to use.[26] For these fifteen studies, they distinguished between categories of antisocial behavior (criminal offending, aggression, and antisocial personality) and between severe and nonsevere manifestations of these. In the severe category they include aggression, as measured either by criminal conviction or reported physical aggression on others, and antisocial behavior, as measured by destruction of property or diagnosis of APD (antisocial personality disorder) or CCD (childhood conduct disorder). In the nonsevere category they include noncompliance with authorities, teasing others, or attacking inanimate objects (e.g., a doll in the psychologist's office).[27] Their finding was that stronger genetic effects could be found for severe antisocial behavior (just under .50) than for nonsevere (.00).[28] Even after all the pruning and distinguishing, there were still methodological problems in the studies. For example, twin studies showed a greater heritability effect than did adoption studies, but only two of them controlled for contact between members of the twin pairs. Moreover, Mason and Frick's attempts to disaggregate the behaviors were thwarted by researchers who treat as a single category "minor offenses sanctioned by fines, felonies, and juvenile offenses after age 15." Although some of the studies reportedly did separate heritability estimates for the different behaviors they included, not all did, nor did Mason and Frick indicate how they extracted information for their estimations for severe and nonsevere forms of antisocial behavior.

---

25. Mason and Frick 1994, 302.

26. This rather catastrophic reduction is surely part of the news, but while acknowledged as significant it does not deter Mason and Frick from trying to extract some hope for heritability.

27. The multiple criteria for APD show how problematic the intuitively plausible distinctions Mason and Frick make are in practice.

28. Like many researchers in this field, they tend to use "heritable" and "genetic" interchangeably. For criticism of such assumed equivalence, see Billings, Beckwith, and Alper 1992.

Neither of these reviews, then, succeeds in reducing the conceptualizations or operationalizations of aggression sufficiently to engage in meaningful comparisons against rival research approaches. Albert's group focuses on the relevance of animal models, but the category they deemed most relevant, defensive aggression, can be operationalized in many ways. The Mason and Frick study tries to sort through and categorize what type of aggressive behavior the research they examined had studied, yet the categorization fails to be fine enough. In the face of such multiplicity, can there be a scientific understanding of aggression? And if so, what could it be? To repeat a point made earlier, without establishing stability across criteria, test, and measurement method, there is no guarantee that any two studies are studies of the same phenomenon, but that must be the premise both of mutual confirmation or agreement within approaches and of critical debate among approaches. These issues are faced fairly directly by Rhee and Waldman in their meta-analysis of behavioral genetic studies of antisocial behavior.[29] They take account of differences in general category and mode of assessment of antisocial behavior, as well as of differences in results from twin studies and adoption studies. They are sensitive to the moderating effects and possible interactions of assessment method and category of antisocial behavior and attempt to control for those. Nevertheless, they leave the categories of APD and CCD, with their menu-style diagnostic criteria, unanalyzed.[30]

What Is (Homo- or Hetero-) Sexual Behavior?

In the previous section, I drew attention to two aspects of the conceptualization of aggression as an object of study. First, only some kinds of aggression are studied, individual non-state-sponsored infliction of harm by bodily, including vocal, means. Second, that category of aggression is operationalized by quite different indices, raising the question of whether there is a unitary phenomenon that is the subject of investigation. The issues in the definition of sexual behaviors are somewhat different. Here the complaint is against the dominance of one measure, the so-called Kinsey scale, which rates individuals on a seven-point scale, from 0 for those who engage in exclusively heterosexual acts and fantasies to 6 for those who engage in exclusively homosexual acts and fantasies. The

---

29. Rhee and Waldman 2002.

30. They also note that the results about which they feel confident suggest the need for better-designed studies, especially longitudinal studies and studies that can distinguish among different types of environmental influence—a need articulated above and partially answered by work like that of Caspi and Moffitt or Kendler, both discussed in chapter 6.

advantage of the Kinsey scale is that, as a one-dimensional continuum, it permits assigning members of a population to at least three groups—homosexuals, heterosexuals, and in the middle, bisexuals—and more if desired. This is an improvement on an earlier dichotomous model, in the view of some sex researchers, because it makes room, at least in principle, both for genuine bisexuality and for a greater variety of erotic feeling and experience, as it distinguishes between fantasies and acts.

Much demographic research seeking to determine the relative frequency of homosexuality uses either a dichotomous standard or the Kinsey scale.[31] Etiological research has tended to concentrate on the production of the extremes. But researchers in sexology, anthropology, and psychobiology have criticized reliance on the Kinsey scale for several reasons. Some researchers point out that the scale is used as though it differentiated stable and uniform human types. When the types in question are at the two extremes of the scale, bisexuality disappears from view. Furthermore, the presupposition of stable human types, they argue, fails to take into account the varying sexual economies of different societies, especially with regard to differences in the ways sexual development is culturally organized and the ways homosexuality and heterosexuality are folded into differentially organized developmental trajectories.[32] Some societies, for example, include a period of obligatory homosexuality in maturation (for males). Others have institutionalized a "third" sex. A Kinsey score must be understood differently for members of those societies than it would be for members of a society that has no such obligatory period, has a different pattern of gender roles and identities, or is characterized by non-obligatory fluidity in sexual behavior and erotic orientation.[33]

Other researchers argue that the scale isolates sexual orientation from other phenomena to which it is related. John Money regrets the use of the term *homosexual*, as directed too much at acts as distinguished from affectional disposition. He proposes the term *homophilia* as a more accurate designation.[34] He further proposes a general category of "gender transposition" with three grades: total

---

31. See review articles by Bailey and Zucker (1995) and Diamond (1993).

32. Herdt and Stoller 1990.

33. Of course, one might defend the scale by pointing out that it could be used in longitudinal studies to determine how individuals in a society with age-structured homosexuality pass through the developmental stages. But in practice the scale is not used in this way.

34. Money 1990. One consequence of this proposal would be the differentiation of the elective or self-endorsed homosexuality of gay men and lesbians in contemporary societies from the practices and identities of those in cultures where homosexuality is a phase of maturation.

(to which he assigns transvestism), partial unlimited (to which he assigns what he calls andromimesis and gynemimesis), and partial limited (to which he assigns what he calls homophilia). Each of these has an episodic as well as continuous mode. While Money worries that critics might object that this posits homosexuality or homophilia as abnormal, it is as likely that critics would object to incorporating sexual or philio-erotic orientation into gender, as gender transposition. Whalen, Geary, and Johnson, by contrast, want to incorporate hetero- and homosexuality into a broader set of orthogonal dimensions of sexual variation, dimensions that would include degree of arousability, frequency of sexual interaction, and number of partners, as well as more finely grained aspects of partner and activity preference.[35] Clearly a typology that captures all of these distinctions will pose a much greater challenge for research design than does the Kinsey scale.[36]

Finally, measurement of sexual preference is challenged by the continuing cultural proscriptions against homosexuality and other departures from normatively sanctioned sexual behavior. Guarantees of confidentiality presuppose both trust and self-transparency, neither of which is probable under the circumstances.[37]

•

Both aggression and sexual orientation, then, turn out to be more difficult to define clearly and carefully enough to support unambiguous studies than one might at first expect. The reasons for and nature of the difficulties are different in the two cases. In the case of aggression, the fragmenting effect of multiple indices and measuring systems can be mitigated by studies that investigate cross-criteria stability. Yet this leaves untouched the complaint that only a subset of aggressive behaviors is the object of investigation. If some aggressive dispositions find outlets in socially approved contexts, and thus remain uncounted, it is not clear about what phenomenon the research is informative. Perhaps the behavior actually measured represents some kind of underlying incapacity or pathology, and not the more specific (or more general) phenomenon that the

---

35. Whalen, Geary, and Johnson 1990.

36. On the other hand, it is conceivable that more finely grained classifications might permit the discovery of tighter links between a given type of behavior and some of its causal antecedents. There is as yet, however, no agreement as to what typological system ought to replace the Kinsey scale. See McWhirter, Sanders, and Reinisch 1990a.

37. See also Diamond 2004 for data indicating the instability of self-report over time.

research promises to help explain. Moreover, a persistent gap between self- and other reporting continues to cast doubt on the validity of constructs measured via such reports. In the case of sexuality, further thought and research is needed to determine if the bi- or tri- (or septi-) valent schema of the Kinsey scale is too coarse-grained or too culturally specific to enable useful investigation. In part, this involves decisions about what one wants to know and why. This question will be addressed in a different context in the next chapter.

Philosophical Analyses and Definitions of Behavior

In addition to the difficulties of defining and operationalizing behavior for scientific research, the colloquial concept of behavior is itself not uniform across the various contexts of its use. Neither the kinds of phenomena that are categorized as behavior nor the definitions of "behavior" are settled. There is a range of meanings, from characteristic patterns of activity (as in the behavior of electrons in an excited state), to function, to something more like intentional action, as in the injunction "behave yourself" or the manner, boorish or gracious, of a dinner host. And one may be speaking of an established pattern or of a single instance. This colloquial instability means that it is not immediately clear what a scientific inquiry into behavior is actually investigating. Examination of the research reveals that it is dispositions to behave in a particular way and not token instances of behavior that are the object of study. The restriction to dispositions does not, however, eliminate the vagueness and ambiguity of "behavior."

Philosophers have identified further different possible meanings of "behavior." They have largely been motivated by traditional philosophical worries about the relation between mind and body and between human freedom and causal determination. Most have been concerned with the ontological commitments carried by the varying explanations of behavior and thus with the possible explanatory commitments entailed by the varying definitions or characterizations of behavior. If behavior is just bodily movement, the consequences for mind and body and freedom and determinism will go in one direction. If behavior is more than that, the puzzles persist.[38] A number of philosophers have sought descriptions or definitions of behavior that did not in and of themselves imply intentionality or nonmaterial causality. Some of this philosophical thinking about the

---

38. These will be familiar to readers of reviews by philosophers of work by neuroscientists and the neuroscientists' replies. See, for recent examples, McGinn 2011; Ramachandra and McGinn 2011; Searle 2011.

concept of behavior can be recruited to help clarify issues in the interpretation of behavioral research.

In *Explaining Behavior*, Fred Dretske attempts to characterize behavior and its causes in such a way as to enable us to make sense of the concept of reasons in a materialist metaphysics.[39] As a prolegomenon to his own analysis, he sifts through a variety of candidates for the denotation of "behavior." In a behavioral episode, say, depressing a lever, what qualifies as behavior is neither the cause of the event (whether the cause is an external stimulus, such as a light flashing, or an internal event, such as perception of the light flashing or some other neural excitation) nor the effect (the lever's being depressed or a food pellet's being released). Behavior, Dretske says, is what something—a person, animal, plant, or even machine—*does*. Neither the cause alone, nor the effect alone, behavior is the cause's *causing* the effect. It is "a complex causal process, or structure, wherein certain conditions or events (C) produce certain external movements or changes (M)."[40] Different causings within the causal sequence also count as behavior. If M, a movement described as the depression of a lever, brings about N, the release of a food pellet, and C is the pigeon's pecking that depressed the lever, then both M's causing N and C's causing N are properly referred to as behavior. "The depression of the lever released the food pellet" and "the pigeon released the food pellet" are both correct descriptions of the same behavior. A given behavior or episode, then, may be broadly (releasing the pellet) or narrowly (depressing the lever) construed, and abstractly (depressing) or concretely (pecking) described. Each way of construing or referring, however, picks out a complex process constituted of an internal state or event that is part of, and in some sense initiates, the process (e.g., some representational or conative state) and an accomplishment (or attempted accomplishment) that is the outcome of the process.

Dretske also distinguishes two kinds of cause about which one might inquire. A *triggering cause* is an internal or external event that directly brings about an effect.[41] Such a cause is an element in a process that has a structure. A typical structure might be E (light flashing) causes I (neural event corresponding to {lever → food}) causes M (the pigeon pecks the lever). Here E is the triggering

---

39. Dretske 1988.

40. Dretske 1988, 21.

41. This is not meant to suggest that there are no other factors (e.g., a state of health) involved but to indicate the cause of an episode rather than of a pattern.

cause that has as its effect I, which is the initial event in the behavioral process {I → M}. A *structuring cause,* on the other hand, is either the set of background conditions that enable one thing to cause another or an earlier event or condition that brought about the background conditions. Structuring causes, so under-stood, are not general conditions (like the presence of oxygen, or good health) that permit any of a family of processes to occur. They are instead specific to the kind of behavioral process in question. The structuring cause for the process just described, E → I → M, is presumably a conditioning regime undergone by the pigeon through which it has learned to associate a light flashing and its pecking a key with the production of food, and the persistence of that association through (at least) this instance of the behavior. To the extent that the approaches exam-ined in this book are concerned with dispositions to behave in certain ways, then the differences among them may be best understood as concerning structuring causes, not triggering causes. Using Dretske's language, the causal question with respect to a behavior expressed by one or more individuals in a population con-cerns what brings about the conditions in which a given kind of event serves as a triggering cause, initiating a process, the organism's or mechanism's moving in response to some internal state in a way that results in M. To put it another way, the question is about the differential inculcation of dispositions: why some individuals in a population are characterized by certain kinds of dispositions and others not.

Dretske offers one further refinement in connection with his primary con-cern, the concept of a reason. Reasons, he says, explain *why* a particular effect N was produced, not *how* it was produced. The *how* includes the particular move-ments used to produce the effect, for example, running as opposed to swim-ming, touching a lever with a proboscis (nose or beak) as opposed to a limb (paw or claw). Any permutation of the latter movements might be identified as (part of) the same behavior—depressing the lever—but they involve differ-ent parts of the neuromuscular system. Reasons explain behavior, not motor patterns. Behavior is the causing of external changes via some motor pattern or other, but it is not identical with or reducible to the motor pattern. A pigeon can learn to touch a key with beak or with claw; the two movements involve different muscles and nerves but accomplish the same end, releasing the food pellet. Releasing the food pellet is not identical to any single movement type. Stereotyped motor behaviors, then, are neither a good model for more flexible animal behaviors nor, *a fortiori,* for human behaviors. Of course, in some cases, for some mechanisms, there may be only one movement pattern that is the motor element of the behavior. A thermostat usually works in only one way, but its temperature regulative behavior is, nevertheless, more than the movement of its springs and bimetallic strips.

Berent Enç takes this point further.[42] Enç is one of the few philosophers to focus directly on behavior as the subject matter of scientific psychology, as distinct from an element in traditional philosophical puzzles. He notes that there has been debate in the philosophical literature concerning whether behavior should be understood "in terms of the effects bodily movement have on the environment" or as "limb movements . . . 'close in' to the body." The difference is the difference between (1) turning on a light and (2) raising one's arm and moving one's hand and fingers. The second of these kinds of description is preferred especially by philosophers committed to materialism or to psychological internalism.[43] Enç quotes Jaegwon Kim on what he called "basic actions" as an example of this preference. Kim was looking for the minimal unit that could be labeled an action—the most basic constituent of complex actions such as paying bills or flipping switches that still warranted the label *action* rather than, say, *muscle contraction.* Enç clarifies the notion Kim was advancing by defining "basic" or "molar" units of behavior as those behaviors "for the accomplishment of which no cognitively controlled behavior is a means." That is, the basic or molar unit is cognitively controlled, but none of its component behaviors or means is so controlled. Kim thought that such basic actions were (1) a minimal unit of behavior and (2) local to the organism, that is, not requiring reference to anything other than the parts of the organism, such as its limbs, for their description. Although B. F. Skinner, the preeminent advocate of empirical behaviorism, did not commit himself on this matter, it is safe to infer from what he did say that he thought the same.

Enç argues, contrary to the minimalists, that the units of behavior that are or should be the object of psychological explanation are not or not necessarily or always fully coextensive with and limited to movements of the organism. The environment in which the movements are executed is relevant to determining just what behaviors they are, and thus their proper description will include reference to something external, like the steepness of a slope or the stickiness of a surface. Molar units of behavior are the behaviors they are because of their functional

---

42. Enç 1995.

43. This is not to say that these two isms are equivalent, just that both require a minimalist description of behavior.

role for the organism, and because organisms behave in environments, specifying this role involves reference to elements in the organism's environment.[44]

While Dretske and Enç have different philosophical aims, they offer similar lessons in regard to the issues concerning empirical behavioral research. Dretske's distinction between the *how* and *why* of behavior suggests a restriction of the concept to teleological processes, that is, processes that have a particular outcome as their aim or as their typical result. If researchers were to take this notion as their basis then the behavioral indices whose occurrence they measure would need to be identified by aim or typical result for the individual. This would likely lead to even greater fragmentation of the construct under study. Alternatively, omitting aim or typical outcome from the definition would increase the gap between human behaviors in their natural settings and the versions susceptible to systematic investigation.

Where Dretske emphasizes the role of outcome in identifying behavior, Enç emphasizes environmental context. If we accept Enç's view, acts classifiable as aggressive behavior need to have targets and settings and cannot be mere movements of the body or limbs, described independently of their environmental target or setting. Lists of the "behaviors" upon which genetic influence is said to be established or studied include alcohol consumption, aggressive and antisocial behavior, and sexual behavior, but also Huntington's disease, Duchesne muscular dystrophy, and Alzheimer's disease (as well as bipolar and other psychiatric disorders). While these latter maladies certainly involve what we might call whole body movement, as do aggression or sexual behavior, that movement proceeds from a physiological or anatomical disorder—destruction or withering of neural or muscular structures—and does not require reference to an environment, aim, or typical outcome. If a researcher is just interested in genes or neural structures, the amalgamation of these phenomena under one umbrella—phenomena of bodily motion, perhaps—is not of much consequence. If the matter of interest is what an organism does, then it does matter. A claim that a given human behavior is largely influenced by genes, by parental rearing, or by some other factor should be assessed quite differently if based on studies of physiologically identifiable disease and mental disorder than if based on studies of "normal" human functioning. To treat pathological conditions as analogues for other behaviors of

---

44. Enç's argument appeals to (1) a variety of animal experiments in psychology showing the difference between conditioning animals to produce a species-typical behavior relative to a reinforcer under certain stimuli, and conditioning them to produce a non-species-typical behavior related to the reinforcer, and (2) analysis of experimental work that brings out the dependence of classification of a behavior on its evolutionary history, and hence its function.

interest is to lay a basis for treating those behaviors as also pathological. While most researchers are sensitive to these distinctions, they tend to get lost as the research is communicated through varying channels—first, the original research report, then review articles, science magazines, and on to mass-circulation magazines or newspapers.

The increase in imputed generality that occurs in opinion pieces or reviews of research is similar to the conflation of operational indices of a particular behavior with the behavior itself that afflicts the interpretation of studies of aggression. Both are instances of what we might term an "upward slippage of significance."[45] Resisting such slippage would disqualify movements of Huntington's chorea as aggression, even if they resembled aggression or had, incidentally, the same effects as an aggressive act. I suspect that the movements labeled hostile or aggressive that are performed by some of those afflicted with Alzheimer's disease may be more like the involuntary movements of chorea than an expression of malice or anger. I'm not sure enough is known about the phenomenology of either Alzheimer's or Huntington's to make solid pronouncements here, but if the movements in question turn out to be a matter of neural failure rather than cognitive/affective reorientation, they just don't belong in the same category as the behavior of the schoolyard bully, back-alley thug, or spousal abuser.

Enç's analysis can also be used to raise questions about the value of certain animal experiments in elucidating the nature and etiology of human sexual behavior. That stressing a gene in drosophila can elicit movements that are components of copulative behavior in that species does not necessarily show that those movements in that experimental context are equivalent to copulative or sexual behavior. It certainly shows that genes related to bodily motion can be affected by environmental factors, but more work would be required to support claims about causes of sexual behavior in that species, and even more to treat this sequence of body movements in the flies as a model of *human* sexual behavior. The animal work Enç cites shows that different "prepackaged" motor sequences can subserve the same function in the same organism and hence count as the same behavior, but also that similar prepackaged motor sequences can subserve different functions and hence count as (parts of) different behaviors in the same or different organisms. Here context is required to distinguish one behavior from another. It doesn't take too much imagination to see that the same sequence of movements could be sexual in one context but not in another, or that some other sequence could constitute aggression in one context but not in another.

---

45. For a notorious and much-quoted example, see Koshland 1989.

Of course, context distinguishes real aggression from play and from acting, offensive from defensive aggression, and so on. To specify just what behavior is in question requires specifying the functionally relevant features of the context. This, in turn, requires interpretation of the situation, making the identification of a behavior a much more complex form of observation than the operationalizations above allow for.

To the extent that Dretske's emphasis on the distinction between the why and the how of behavior identifies the pigeon's reason for pecking the key as its belief that pecking will produce food united with its desire for food, it diminishes the status of reasons as part of the human repertoire. More interestingly, however, it suggests a possible distinction among nonhuman behaviors. Stereotyped motor behaviors like mounting, lordosis, or flank marking among rodents can be induced by interfering with genes or hormones. Can they also be induced by operant conditioning? If the answer is negative, one might be tempted to make this the basis of a distinction between those animal behaviors that can serve as models of human behavior (those susceptible to operant conditioning) and those that cannot (those not modifiable by operant conditioning), or at least to pair animal models to human behaviors type to type. Much of the genetic and hormonal animal research might turn out to be irrelevant to the understanding of human behavior (as distinct from the understanding of some aspect of human genetics and physiology), except for those behaviors that are a result of a genic or hormonal or neurophysiological abnormality.

Both the Dretske and the Enç proposals suggest caution in interpreting the various studies of human behavior made using operationalizations such as those discussed earlier in this chapter. An operationalization that does not include reference to typical outcomes or to the environment in which a movement is executed is not likely to approximate human behavior in natural settings.[46] It may be too broad to capture a genuine human behavior, or it may be too narrow to serve as a surrogate for the behavior the study hopes to illuminate.

In an essay concerned with the prospects of research on the genetics of human violence and aggression, Allan Gibbard makes some further distinctions.[47] He approaches the topic from the perspective of evolutionary psychology, but his emphasis is on the broad array of causal pathways that might have incidents of violence (or, by extension, any other type of behavior) as their outcome.

---

46. For further discussion of the gap between operationalizations of behavioral constructs and the behaviors these are intended to illuminate, see Longino 2002.

47. Gibbard 2001.

These pathways parallel the approaches analyzed in previous chapters and range from genetic mutations that might result in an individual's expressing some behavior B with greater frequency than others in her or his society, to environmental conditions that enhance the survival of those inclined to behavior B relative to those not so inclined, to hormonal configurations inhibiting or increasing a given behavior, to environmental conditions modulating such hormonal configurations, with many other possibilities in between. In painting this canvas, Gibbard is making a primary distinction between two kinds of explananda: in the case of violence, individual acts (of violence) and "conditional plans for violence." A conditional plan might be "If conspecific of same sex approaches self's mate and is smaller than self, attack conspecific" or "If conspecific of same sex approaches self's mate and is larger than self, flee." An individual with such a behavioral program might never fight or flee if the triggering environmental conditions are never realized. Evolutionary psychology, Gibbard emphasizes, is about the evolution of such plans or programs (or dispositions to perform such processes), not the evolution of the behaviors themselves. Behavioral genetics, by contrast, is about the heritability of behavioral traits in a population in an environment and uses as evidence measurable instances of the behavior (along with presumed kinship facts concerning those whose behavior is measured). It uses as evidence, and is about, measurable differences in the expression of a trait among individuals in a population, and not about internal psychological states. Since the behavioral plans evolutionary psychology studies are (if they are indeed adaptations) species characteristics, (1) there will be no difference among individual members of a population in possession of these plans for behavioral genetics to study,[48] and (2) any differences in the expression of behaviors specified by those plans will be the outcome of environmental, not genetic differences.

Gibbard's punch line is that behavioral genetics (by which he means quantitative behavioral genetics—the study of heritability) is of little use in understanding the problems of violence that afflict some contemporary societies, such as the United States. The relevant question is not why some Americans are violent and others not, though this is undoubtedly the case and certainly a situation that could be investigated, but rather why the level of interpersonal violence in the United States is higher than the level of interpersonal violence in, say, Sweden or Italy, and if and why different types of violent act might predominate in these

---

48. Here either Gibbard or his sources overlook the possibility that natural selection might have favored a particular nonuniversal distribution of a conditional plan within a population. Indeed, much of evolutionary psychology concerns putative sex-specific behaviors (or behavioral plans).

different settings. Since the settings and these differences are far too recent to be the result of natural selection, evolutionary psychology would also be of little relevance here.

Gibbard's discussion reveals at least three kinds of question that might be asked by a scientific research program concerned with accounting for a particular behavior. One question asks how and to what extent variation in a population can be explained. A second asks how and to what extent a trait common to a population can be explained. As he points out, behavioral genetics is concerned with the first of these, evolutionary psychology with the second. One studies heritability, while the other studies the extent to which behavioral traits are inherited. The distinction to which Gibbard draws attention is not always observed. This may be because both *heritable* and *inherited* are commonly contrasted to *learned*, which may account for their not infrequent conflation.

Gibbard also draws attention to a question concerned with a third kind of behavioral phenomenon. Contemporary populations differ from each other in the frequency with which various kinds of behavior are expressed. What accounts for such differences? Here neither methods designed to account for differences among individuals (behavioral genetics or family or developmental psychology) nor those designed to account for differences among species (evolutionary psychology) are of any help, the former because their findings can only be applied within populations, the latter because the populations we are concerned with are of the same species. Species differences, or adaptations, long antedate (because of the time it takes to fix a trait) the formation of contemporary human populations (themselves products of much intraspecific mixing). This third behavioral phenomenon is about properties of populations and cannot be studied with methods designed to study either properties of individual organisms or differences and similarities among members of a population.

.

These philosophical reflections raise the question of what should be included in an instance or unit of measurable behavior, that is, what is the proper ontology of behavior. Most of the research approaches surveyed earlier in this chapter have shared one major supposition about behavior, namely, that behavior is a property of individuals. On this assumption, to understand behavioral phenomena it is necessary to investigate the etiology of individuals' dispositions. A few of the studies mentioned, however, have hinted at different ways of thinking about the ontology of behavior. The studies by Lavigueur and by Speltz looked at interactions in parent-child and parent-parent dyads.[49] It is possible to see those interactions as properties of the dyads, or as relations between the members of a dyadic (or triadic or *n*-adic) complex, rather than as properties of individuals.

To think in terms of interactions and relations is to move up a level of organization. Many thinkers about animal behavior have supposed that the alternative to thinking individualistically is thinking collectively.[50] The dichotomy is then between an ontology of individuals and one of groups. But to think about behavior relationally is to pose an intermediate ontological level: that of a small group, numbering from two to four or five members, that is part of a larger social group. While the membership of groups can change, what is key is that a behavior is the behavior it is in the context of reactions and interactions by and among members of such small groupings. The same movement may be part of an aggressive interaction in one context but part of a peaceful interaction in another. Two ontological possibilities are opened up here. In those instances in which it makes sense to think of behavior as attaching to only one individual, the behavior is the behavior it is because of the context in which the constituent movements are performed. In other instances, the behavior may need to be understood as performed by two or more individuals, that is, it may require conceptualization as *interaction*, not as action.

Primatologist Frans de Waal has called explicitly for such a relational approach to aggression, and to social behavior more generally.[51] He is less concerned with the ontology of behavior than with emphasizing that primates, including humans, are social animals, always in networks of relationships. Their behavior occurs in such networks and relationships and has the character and function it does because of the particular network and relationship within which it occurs. For him, aggression is just part of a species' behavioral repertoire and is integrated with other behaviors to constitute the fabric of the species' social life.

In either case, what must be measured is movement within a specified kind of context, or interaction among two or more individuals. These alternative behavioral ontologies lend themselves much better to Gibbard's third kind of

---

49. Lavigueur, Tremblay, and Saucier 1995; Speltz et al. 1995, discussed in chapter 3. These are just two of many examples of a research focus on interactions.

50. See, for example, de Waal (1989, 24): "There are globally three perspectives from which behavior can be studied . . . the group as a whole, . . . the individual, or . . . the genetic material."

51. This stands in some tension with the tripartite division just noted. De Waal might well reconcile the two positions by saying that the behavior of individuals must be studied relationally, as interaction.

question, the question of why two different societies differ in the frequency with which a given behavior is exhibited or in its distribution.[52] Different populations of other primate species have also been shown to vary in the frequency of aggressive behavior. What would account for different behavioral patterns when the gene frequencies are the same? Similar questions are generated by anthropological work on sexual orientation. Why does the timing and frequency of same-sex sexual interaction vary among different human societies? How and why has the distribution of same-sex sexual interaction changed in a given society over a specified period of time? Here the focus is on the different frequencies of varieties of sexual expression in different populations.[53] Behavior as an object of investigation is thus treated as a population-level property, rather than individual-level property. The questions to be investigated concern the differences among different populations in the frequency and/or distribution of behaviors. These are questions of the kind addressed by the ethological and ecological approaches described in chapter 7.

Combining an interactionist and contextual conception with a population conception of behavior leads to a very different ontology than the individualist ontology that characterizes the approaches embroiled in the so-called nature-nurture debate. In either an interactionist or a population conception, the object of measurement—interactions and frequencies and distributions of interactions—is different from the object of measurement when one is focused on individual variation. The questions of the nature-nurture debate represent only a subset of the questions we might ask about human behavior.

Conclusion

Human behavior is so familiar to us that it is difficult to achieve sufficient distance to devise stable and meaningful constructs for investigative purposes. The two families of behavior whose scientific investigation form the subject of this book turn out to be especially elusive. Aggression, even when narrowed

---

52. De Waal (1992) also takes this to be the central question concerning violence generally. For him, aggression is simply one strategy available to all members of a population and can be both adaptive and maladaptive; the interesting question is why different societies manifest different levels. De Waal has argued that bonobo societies differ markedly from other primate societies in the level of aggression (and sexual interaction) they display. More recent work is suggesting that their aggressive and sexual behavior in captivity and in the wild is quite different, thus calling de Waal's specific observations into question.

53. Imagine how much richer the investigation if the questions were opened up to the full range of variation in sexual preference.

to non-state-sponsored interpersonal infliction of harm splinters into different measurable indices. Research attempting to associate frequency of one of those indices with a putatively causal factor can, if successful, suggest that that behavioral index is causally dependent on the factor. Without robust cross-index concordance, however, such a study is limited in its relevance to the particular index studied, not to the larger category. Furthermore, even cross-index concordance does not guarantee that the construct so operationalized is informative about the behavior in its everyday transactional context. The measurable indices are held together by a folk understanding of aggression, and while a successful study can provide partial knowledge of some subcategory or subpopulation, it is limited in scope. The focus on sexual orientation, on the other hand, may conceal other dimensions of sexual variation that interact with the object of erotic attention/arousal. Here the partiality results from too coarse a measuring instrument, rather than from a multiplicity of indices.

A further difficulty with the indices concerns not their variety but their ability to encompass the phenomena in a way that permits a full understanding of their etiologies. Not only are the causal factors differently parsed, but the phenomena to be explained are also susceptible to different parsings, with different criteria of identification and individuation. What behavior is, what is to be measured in any given approach, is not given in the phenomena but is a function of what kind of question researchers seek to answer. As suggested earlier, where questions about gene expression or physiology drive the inquiry, genetic and physiological correlates of behavioral differences and commonalities are a clue to further research into the workings of the organism, not sought necessarily as explanations of the behaviors. Where behavior is the focus, philosophical analysis of the concept of behavior points to the necessity of exercising greater care in selecting what to measure and how to measure it. The question about the proper ontology or units of behavior does not have a unique, universal answer. The proper ontology depends on the question being asked. The questions are many and reflect a diversity of cognitive and practical interests.

10

The Social Life of Behavioral Science

P revious chapters have analyzed the evidential and conceptual aspects of behavioral research. Here we will consider the career of these studies in the public realm. Some commentators assume that there is a direct connection between the research approach represented by the content of articles and political outcomes. This is especially the case when critics focus on just one of the approaches. I will argue that the relation is more subtle and indirect than often supposed by critics, but no less a matter for concern.

As response to the "Genetic Factors and Crime" conference, mentioned in chapter 1, shows, negative critical attention is most often attracted by genetic and biological approaches to behavior. But part of the burden of this book has been to show the variety of approaches in behavioral research. Surely, one might say, that should be sufficient to check the deleterious social impact of any single approach. Furthermore, much research goes unremarked, with little or no impact on policy, practice, or public attitudes. Behavioral research does, however, excite more than the usual interest and thus finds its way to the public through the pages of popular science journals, the science sections of newspapers, other mass-media publications, and books written for lay consumption by researchers themselves. Through this interface a picture of the state of scientific understanding is communicated to policy makers and the general public. To get a sense of the impact of the research discussed, it is necessary, then, to see how it is filtered through these channels of knowledge dissemination.

The Circulation of Knowledge

Aided by a number of research assistants, I undertook an analysis of the uptake of selected articles representative of the different approaches. We studied citation patterns, the reporting of behavioral research in mass-media and "middle-brow" publications, and the reviews of general-interest books by researchers.[1] The goal was to obtain a more complete picture of the transmission of ideas, both within the research context and between the research context and nonspecialist readers. Only after that is accomplished is it reasonable to talk about the social impact of this work, reflections on which will occupy the second part of the chapter.

Uptake among Peers

With the assistance of Whitney Sundby, I analyzed patterns of citation for several articles or collections of articles mentioned or discussed in earlier chapters.[2] The point of this analysis was to see how the material in those articles was taken up by other researchers. We used both the ISI Science Citation Index and the citation record available through Google Scholar. Although both books and journals are included in citation information, the following analyses are restricted to journals. Two kinds of comparisons are of interest: (1) relative quantity of citations and (2) distribution of citations in different kinds of publication. After reviewing several sets of citations and the mastheads of journals included in those sets, we came up with eight categories of journal: research, clinical, clinical/research, policy, policy/research, policy/clinical, policy/clinical/research, general interest, and miscellaneous.[3] Abstracts and articles themselves inform about

---

1. This empirical part of the project was the most challenging for a philosopher to perform, and I benefited from the help of a series of research assistants. Yeonbo Jeong, Kristen Houlton, Cale Basaraba, Mario Silva, and Whitney Sundby all contributed to this effort, tracking down and organizing large quantities of information. Information technology improved immensely through the lifetime of this part of the project. Laborious searches in the library have been replaced by finger-tip commands on computers linked to the libraries' databases, electronic (and digitized) journals, or Google Scholar. While the analysis could certainly be more extensive and deeper, this would require resources, both of time and of person-power, beyond those available to me. Nevertheless, I think this preliminary survey is sufficient to give an overview of uptake patterns. For an example of a study of popular media representation of science, albeit confined to only one of the approaches reviewed in this work, see Nelkin and Lindee 1995.

2. Preliminary work was performed by Kris Houten, Cale Basaraba, and Mario Silva.

3. We defined the categories as follows. *Research*: any journal whose primary focus is research; this includes basic/experimental research and research to inform theory. *Clinical*: any journal dedicated primarily to the discussion, assessment, diagnosis, or treatment of patients. This includes medical journals and journals specifically targeted to practitioners in medical fields. Research is often a part of these journals, since most

the content; the journal category informs about the intended audience for the content. In many cases, articles published in clinical journals or policy journals seem little different in content from those published in research journals. The difference is in the implied expectation that the content is relevant to a readership interested in potential clinical or policy applications of the research. Finally, the number of citations tells us something about the relative impact of the article or set of articles.

The research approach outlined by Caspi and Moffitt has generated a great deal of interest. While others were probably pursuing similar lines, Caspi and Moffitt were successful in attracting attention both within their local research context and among downstream users, as well as in popularizing media such as *Scientific American* and among philosophers. A study of a set of articles citing their work reveals uptake in research, clinical, and policy contexts.[4] Articles linking a specific genetic mutation with a behavior of interest received many more citations that those advocating the interactionist approach with which Caspi and Moffitt are most associated. While a few of the articles citing the 2006 *Nature Genetics* presentation of their GxExN approach reported negative results on gene-behavior associations, most either presented particular studies within the interactive framework showing gene-environment interactions, proposed methodological refinements, or promoted the GxE or GxExN approach, or informed new constituencies of its relevance to their concerns. Table 1 (see appendix) shows the distribution of citations to a set of articles by Caspi, Moffitt, and various coauthors through mid-2009.[5]

The behaviors and conditions to which the interactive approach is applied

clinically oriented journals are at least partially based on research. *Clinical/research*: journals that publish both clinical papers and research papers relevant to some specific topic (e.g., *Journal of Personality Disorders*; *Journal of Epilepsy*); journals that give equal weight to basic and clinical research; and journals where the primary content is research but the explicit primary purpose of that research is to inform clinical practice. *Policy*: journals that focus on policy, law, or justice. *Policy/clinical/research*: journals dedicated to a broad topic that represents an intersection of research, clinical practice, and policy—where each of these categories seem to be equally relevant. Journals dedicated to social work and services, aggression/maltreatment/abuse/trauma, substance abuse and addiction, public health and preventive medicine, etc. often fall under this category. *Policy/clinical*: similar to the above but with less emphasis on research. *Policy/research*: similar to the above but with less emphasis on clinical practice. *Miscellaneous*: journals that focus on theory, philosophy, research in totally unrelated fields, etc. *General interest*: nonprofessional or mass-circulation media.

4. The articles include Caspi, McClay, at al. 2002, with 2,513 citations; Caspi, Sugden, et al. 2003, with 4,205 citations; Caspi, Moffitt, et al. 2005, with 446 citations; and Caspi and Moffitt 2006, with 535 citations, as well as those used to construct table 1 below. (Citation numbers as of May 2012.)

5. Based on Moffitt, Caspi, Kim-Cohen, and Taylor 2004; Moffitt 2005a, 2005b; Moffitt, Caspi, and Rutter 2005.

in articles citing the Caspi and Moffitt work include depression, anxiety, PTSD, phobias, self-injury and suicide, delinquency, ADHD, substance abuse (including illicit drugs, alcohol, and nicotine), eating disorders, bipolar disorder, schizophrenia, antisocial behavior, and aggression. The model is also proposed as useful in addressing a variety of stress-related somatic conditions, such as chronic fatigue syndrome, fibromyalgia, and irritable bowel syndrome. In animal models it is used to study premature aging, stress-coping, mania, and immune system function. A certain amount of research is going into identifying intermediate conditions (endophenotypes) in the pathway from genes to behavior, such as impulsivity and behavioral inhibition. In addition, the research concerns proteins whose production is affected by relevant genes and particular brain and nervous system conditions. Multiple genes are being studied, as are multiple environmental factors. Methodological refinements include computational models to overcome problems of small sample sizes, definitional and classification refinements, the addition of gender and race as factors, neural models for the action of specific molecular/genetic configurations, and methods of taking advantage of so-called "natural experiments." Reviews of the research were published in journals of public health and health policy and in journals specializing in relations between behavioral science and the law.[6] And the Office of Behavioral and Social Science Research at the National Institutes of Health announced in 2008 a plan to support interdisciplinary and integrative research into behavioral problems.[7]

By contrast, the articles citing Gilbert Gottlieb's work are concentrated in research journals and show a four-to-one ratio of theoretical to empirical content. We chose two Gottlieb articles and two books for our analysis. The theoretical articles citing any of these works include proposals for understanding the interaction of genetic, physiological, and environmental factors in development, as well as calls to incorporate developmental research into evolutionary research and into evolutionary psychology. Empirical articles were concerned with endocrine secretion and behavioral development, maternal metabolism during gestation and offspring intelligence, imitative learning in chimps, the timing of walking in humans, and the reinforcing effect of experience in neonates. The plurality of citations in the research category occurred in the journal *Behavioral and Brain Sciences*, found in five target articles and their commentaries. Unlike work by

---

6. *Criminal Science and the Law, Forensic Psychology, Psychology, Public Policy and the Law, Psychological Science in the Public Interest, Social Policy and Society, American Journal of Public Health,* and other epidemiology and public health journals.

7. Mabry et al. 2008.

any of the other authors reported on, Gottlieb's work also received a significant number of citations in philosophy journals (category: miscellaneous). Table 2 (see appendix) shows the distribution of citations to the Gottlieb work.[8]

Coccaro's work on serotonin shows a pattern similar to that for the Caspi and Moffitt set. While some of his publications get upward of five hundred citations, we chose a set with a more tractable number to study.[9] As table 3 (see appendix) shows, of 139 journal citations, 34.5% were in research journals, 12% in clinical journals, and 43% in clinical/research journals; 9% were in one or another policy category, 5 (or 45%) of those in law or criminology journals.

Looking at single articles rather than compilations reveals further patterns among citations of behavioral research. A 1994 article by Goldsmith and Gottesman, making a case for the behavioral genetic perspective on antisocial behavior, received just under 60% of its citations in research journals.[10] Of these the majority were empirical studies bearing out one or another of Goldsmith and Gottesman's suggestions (re the important, but not determining, effect of genetics on antisocial behavior), either on effects of common rearing versus genetics/ heritability or on particular behavioral traits. Of the articles we classified as theoretical, most focused on methodological issues of measurement or classification, rather than proposing new models or refinements thereof. There were a number of citations in clinically oriented journals. Interestingly, these tended to discuss familial (biological) correlations without distinguishing between genetic and environmental contribution.

A more recent article, a 2003 meta-analysis of behavioral genetic research on aggression by Soo Yun Rhee and Irwin Waldman, makes similar but updated points.[11] It received far more citations than those for Goldsmith and Gottesman, and distributed somewhat differently. Citations in research journals account for just under 40%, and there are almost as many citations in journals classifiable as jointly research and clinical. While the policy and clinical uptake of the Gottesman and Goldsmith work is negligible, at least 12% of the Rhee and Waldman citations are in policy-related journals, and over 36% are in clinical and research journals (compared to just under 15% for Gottesman and Goldsmith). In addition, at least 7% of citations are in journals with an explicit focus on criminology. This is partly accounted for by the fact that Rhee and Waldman presented a ver-

---

8. Gottlieb 1991, 1997, 2001b; Gottlieb, Wahlsten, and Lickliter 2006.

9. Coccaro, Kavoussi, and Lesser 1992; Lee and Coccaro 2001; Best, Williams, and Coccaro 2002.

10. Gottesman and Goldsmith 1994.

11. Rhee and Waldman 2002.

sion of their paper at a criminology conference. Both the citations and the conference presentation are evidence of a shift in interest toward genetic analyses in the criminology literature.

Cathy Widom's 1989 article on the effect of abuse of children on their later offending showed a yet different pattern of citations. Less than 20% of citations were in journals dedicated only to research, with another 20% in journals that published research related to clinical practice or to policy. The highest percentage (39.7%) were in journals that included work on policy, clinical application, and relevant basic research. Citations in these combined policy categories were divided among law and criminology journals, public health journals, and social work journals.

Several features stand out from a perusal of the content of this batch of citing articles. We can distinguish between within-approach uptake, reflected in citations in articles that follow or endorse the approach of the cited article (e.g., behavioral genetics articles citing other behavioral genetics articles), and cross-approach uptake, reflected in citations in articles reflecting a different approach than that of the cited article (e.g., behavior genetics articles citing social-environment articles). Within the research category there is very little cross-approach interaction. Exceptions, such as Gottesman and DiLalla's reply to Widom, prove the rule. Most within approach research uptake concerned ways to extend the approach associated with the cited article or reports on research following a similar interpretative approach. Both genetic and neurophysiological approaches received a significant number of citations in journals oriented to a clinical readership, while the developmental systems approach remained in the context of research journals, with little uptake in either policy or clinical journals. While the number of citations in policy-oriented journals remained negligible through 2005, it should be noted that a number of edited collections or textbooks advancing aspects of a biological point of view (genetic, neurophysiological, "biosocial") have appeared in the years since.[12]

When we considered citations to the population-level researchers discussed in chapter 7, we found both similarities to and differences from the above patterns. Although Blumstein's early work on what he calls "the criminal career" has been widely cited,[13] his more recent article on the puzzling relation between rising incarceration rates and stable crime rates received only eighteen citations

---

12. Including, but not limited to, G. Anderson 2007, Glicksohn 2002, Walsh and Beaver 2009.
13. Blumstein 1986.

in journals.[14] Most of these were in policy or policy/research journals, with only 11% in research journals. The topics of citing articles included incarceration policy, crime rates (including both homicide and drug offenses, among others), and domestic violence. One or two of the journals are general sociology or public health journals, but most are dedicated to criminology, law, or violence. While the distribution of citations for Rhee and Waldman is reversed, the 7% of their 186 total citations that appeared in criminology journals amounts to thirteen in that category, and the 12% in policy journals represents a total of twenty-one— more than the overall total for Blumstein. By 2009 Fagan and collaborators received two and five citations, respectively, for their 2004 and 2003 articles on the relation between crime rates, arrest rates, and neighborhood characteristics.[15] This population-based, ecological approach receives far less uptake overall than the individually focused etiological studies, whether these concern genetic, environmental, physiological, or integrated etiology.[16] On the other hand, as of August 2011, the Sampson, Raudenbush, and Earls study had received over two thousand citations, but the citing articles concern applications of the collective efficacy model to public health issues, such as diabetes and obesity, racial aspects of crime, social organization, and neighborhood structure.[17] A miniscule percentage concern crime or crime rates.

Barry Adam's work on sexual orientation and population-level factors received a number of citations (twenty-three in journals) comparable to Blumstein's.[18] As it was published more than ten years earlier, however, this does not indicate comparable uptake. Over 70% of the citing articles appeared in research journals, just under 20% in policy/research journals, and just under 9% in miscellaneous. All but two were in journals concerned with gender and/or sexual orientation. A similarly themed article by Mildred Dickemann received nine citations, mostly in articles on sexual behavior among various North American Indian cultures.[19] Just for comparison, Bailey and Pillard's genetic study of male homosexuality received, according to Google Scholar, 406 citations.[20] The first

14. Blumstein 1998.

15. Fagan and Davies 2004; Fagan, West, and Holland, 2003.

16. The figures above still represent citations through 2005.

17. ISI Web of Science, http://apps.webofknowledge.com/ (search term "Sampson, Raudenbush and Earls 1997"; accessed August 16, 2011).

18. Adam 1985.

19. Dickemann 1993.

20. http://scholar.google.com/scholar (search term "Bailey and Pillard 1991"; accessed August 19, 2009).

XQ28 study published by Dean Hamer and his colleagues received 550.[21] Citations to Bailey and Pillard occur primarily in articles on sexual orientation, with a few also on determinants of behavior or personality generally. Most of the citations to Hamer et al. are likewise concentrated in articles about sexual behavior or sexual orientation, with a few on molecular genetic methodology and a few on genetic influence on behavior more generally.

•

Several general trends can be identified here. Research focused on the etiology of behavior conceptualized as an individual phenomenon both predominates in the behavior literature and receives much more attention, insofar as attention can be measured by citations. While research on aggression is taken as relevant to research on other behaviors or behavioral families and to general approaches to the study of behavior, research on sexual orientation is taken up primarily by others researching sexual orientation. This gives reason to think that research on human sexual behavior will remain isolated from research on other aspects of human behavior, getting cited elsewhere for the most part for purposes of adding evidence for a particular theoretical approach.[22] Research on sexual orientation receives somewhat more attention than does research on aggression in nonacademic media, mostly in books or journals concerned with sexual behavior or sexual orientation or with gay culture and politics. Individually focused research receives much more attention in these nonacademic venues than does the population approach associated with Barry Adam. The resulting impression is that the real issues associated with the incidence of aggressive or violent behavior or with sexual orientation concern how individuals become disposed one way or the other, and the research debates concern what factors are most influential in the development of individual dispositions. Population-level questions, for example, about variation across differently situated populations, figure much less prominently in the professional discussion of behavior research.[23] While the Sampson,

---

21. http://scholar.google.com/scholar (search term "Hamer, Hu, Magnuson, and Pattatucci 1993"; accessed August 19, 2009).

22. This represents a different pattern than that found in the literature on animal research, especially ethological, field research, which tends to integrate sexual behavior with other behaviors of the population under study.

23. The above analyses (with the exception of the report on Sampson, Raudenbush, and Earls) are based on citations through mid-2009. For comparison, here are citation citation counts as of mid-2011 for several of the articles representing different approaches discussed earlier. Kendler, Gardner, and Prescott (2002),

Raudenbush, and Earls article has received significant citations, these appear to be concentrated in sociology or community studies journals rather than journals focused on behavior or on criminality. Discussions of intervention or prevention follow patterns similar to the trends in the research literature, with an emphasis on intervention strategies that focus on individuals, rather than on populations. Very rarely do questions about population-level variation appear in discussions of genetic, neurophysiological, or psychological/social-environmental factors in the inculcation of individual dispositions, and then generally in discussions by philosophers, not by the researchers themselves.

Uptake in the General Press

Citation analyses such as the above show how research is taken up by peers. But how does it appear to the general public? What does the general public think are the important scientific questions about human behavior? What does the public think is known about behavior? One way to address this question is to investigate the ways in which the research is represented in the mass or lay press. Here there are three principal sources: articles in the science sections of news-weeklies or daily papers and in science periodicals such as *Scientific American*, books by researchers intended for general rather than professional audiences, and reviews of those books.[24] My assistants and I searched for accounts of research on aggression and violence and on sexual behavior as well as for accounts of the theoretical underpinnings of behavioral research.

AGGRESSION AND VIOLENCE RESEARCH IN MAGAZINES

Between 1990 and 2005, *Time*, *Newsweek*, and *US News and World Report* each published twenty-six articles reporting on research on aggression, crime, and violence. The aggression research and debate most reported in these news-weeklies concerned the effects of media violence, especially in video games and

---

presenting their integrated model of the etiology of depression (which stresses the role of adverse events, without excluding genetic factors), had received 374. Caspi, Sugden, et al. (2003), on the role of a transporter gene on depression, had received 3,302. The Risch et al. (2009) meta-analysis disputing the Caspi group's report had received 370.

24. In this work I was ably assisted by research assistants Kristen Houlton, Yeonbo Jeong, and Cale Basaraba. I selected the journals, set keywords, and reviewed results; the research assistants performed searches in journal table of contents, databases, periodical indexes, etc. We did not study books by science journalists, nor did we consider broadcast (radio and television) science reporting.

on television, on young people. The science section of the *New York Times* in this period featured twenty-eight articles, none of which concerned research on media violence.

Newsweek in this period featured nine articles on media violence, all but one of which focused on its effects on young children. Four articles discussed the effects of drugs or addiction on violence. One featured neuroscientific study of brain regions implicated in aggression. Three reported on other research on etiological factors implicated in childhood violence (one each on daycare, excessive self-esteem, and loss and mourning). The rest were about crime statistics or recent episodes "in the news."

Time also published nine articles on the relation between media violence and youth violence. Of the remaining seventeen, two focused on the possibility of a relation between availability of guns and rates of youth violence, while one reported on the same daycare study as reported in *Newsweek* and one focused on approaching antisocial personality disorder as a neurological disorder. None focused on drugs. Matt Ridley's *Nature via Nurture* was the subject of a cover story, and a report of Caspi and Moffitt's studies on MAOA was headlined "Search for a Murder Gene."

US News and World Report featured eleven articles on youth aggression, crime, and violence, only five of which focused on its relation to media violence. Others reported on efforts to direct young people away from violence or on statistics (interestingly, these were statistics indicating decreases in some forms of youth violence throughout the 1990s). Two articles reviewed the state of research in one or another approach to the study of human aggression. In 1990 evolutionary studies were reviewed, with emphasis on the role of aggression in group violence and social control and the role of genetics and neurotransmitters in the etiology of individual aggression. In 1997 a cover story on "The Politics of Biology" covered claims made on behalf of molecular behavioral genetics, but noted the lack of replication of most such studies and difficulties in defining behavioral phenotypes such as alcoholism. The claim of relevance of some of the research to gay rights was covered and a fuzzy interactionism described in the piece's conclusion. This magazine also published a report on Moffitt's MAOA study in New Zealand, but the headline, unlike that in *Time*, incorporated the interaction message of the study: "Genes + Abuse = Trouble." *US News* also featured a profile of Steven Pinker, as did *Time*, during the marketing period for *The Blank Slate*.

In the *New York Times* science section, articles focusing specifically on aggression were fairly evenly distributed among the psychological, social, neurophysiological, and genetic approaches, with a bit of advantage to the genetic research.

The *Times* even carried a profile of a researcher, Felton Earls, collecting data on the "collective efficacy" approach to neighborhood variation. But the newspaper also featured a profile of Steven Pinker, as spokesperson for the scientific approach to understanding human behavior, as well as several articles about Craig Venter, leader of one of the two teams to sequence the human genome. One of these featured representations of Venter's chromosomes with areas putatively related to phenotypical traits, including behavioral ones.[25]

If one could characterize the overall representation of research on violence and aggression in these mass-circulation media, research on genetics has pride of place. The profiles of Pinker and Ridley emphasize the genetic point of view as the front line of research. Criticism is presented as but a tempering of claims, rather than as representing an alternative framework for research. The interactive model offered by Caspi and Moffitt's work is treated as an insight into how a specific gene works rather than as an avatar for an entirely different approach to thinking about behavior. While research on exposure of children and others to media violence was more frequently reported, the issues at stake often had to do with the effectiveness or desirability of one or another kind of control (parental, state regulation of publication and broadcasting, the V-chip, and others). And much of the research on social-environmental influences on aggression presented in chapter 3 remains invisible.

Different approaches to etiology are treated differently when these lay media discuss what correlational research there is. Where the overall framing of reports of research on behavioral genetics research suggests that future research will provide more knowledge, whether of environmental interactions, of gene regulation, or of single-gene disorders, the framing of reports on exposure to media violence leaves an impression of terminal inconclusiveness. The difference is subtle, but the implication is that genetic research will produce results, while environmental research, at least on media exposure, will not. Furthermore, most reports about research are phrased in terms of direct effects on individuals of the factors under study rather than of distributions within and among populations. And even when population-level factors are reported, they are treated as clues to effects on individuals. The overall effect is to preserve an understanding of behavior as an individual phenomenon and group behavior as an aggregation of individual behaviors.[26]

---

25. Wade 2007.

26. In a recent startling example of the disproportionate credibility given to the genetic approach to aggression and crime, the *New York Times* ran an article about the 2011 National Institute of Justice annual

AGGRESSION AND VIOLENCE IN US SPECIAL-INTEREST PERIODICALS, 1990–2005

The *Economist* included only seventeen articles related to aggression, of which one was an obituary for sociologist Robert K. Merton. Five others concerned crime statistics. Two (in the same issue in August 1993) concerned children's exposure to media violence, and two, about ten years apart, concerned genetics research. In 1992 the magazine carried a generally favorable report on increased research in molecular behavioral genetics, incorporating a caveat on environmental interactions. In 2003 the magazine featured an article on Moffitt and Caspi's New Zealand studies of depression, which, like their work on MAOA, pointed to specific gene-environment interactions. One article reported on brain research, one on alcohol and aggression. The remainder concerned specific incidents in the news.

New Scientist, with twenty-seven articles dealing in some way with aggression and violence, featured seven on media violence. Three articles reported on non-human animal research, either field research or laboratory work. There were also a profile of Adrian Raine and his physiological research, a report on the Caspi and Moffitt study, and articles citing work on bullying and self-esteem and on daycare and aggression in children. Several concerned drugs and violence, both the enhancing and the dampening effects.

As in the mass-circulation magazines, the focus in these media is on individual expressions of violence. Aggression is represented negatively, as synonymous with violence and as a phenomenon to reduce or control.

BOOKS AND BOOK REVIEWS

Researchers attached to each of the covered approaches have written books for nonspecialist audiences in addition to their scientific work. These have been received in the general press as entrants into the nature-nurture debate and assessed for their effectiveness in shoring up one or the other side. Reviews in the scholarly press are more likely to be critical, to review flaws in argumentation, incomplete or one-sided presentations of evidence, and failures to address rel-

---

conference, entitled "Genetic Basis for Crime: A New Look" (P. Cohen 2011). The article featured interviews with researchers such as Terrie Moffitt and Adrian Raine about the resurgence of interest in research on genetic and physiological bases for criminal behavior. A look at the conference program (US Department of Justice 2011) reveals that none of those researchers were speaking and that sessions devoted to genetics were concerned with the use of DNA technology in forensics, not in understanding criminal behavior.

evant issues. These are rarely picked up by reviews in the general press, where authors' own representations of the dialectical status quo tend to pass without modification into reviewers' representations of the background against which to assess a book's significance.[27]

Using the EBSCO search tool, we searched for reviews of seven books published up to 2005. Steven Pinker's *The Blank Slate* received the most attention, with forty-two reviews in media ranging from anthropological and general science journals to middle-brow publications like the *New Yorker, Times Literary Supplement,* and *New York Review of Books,* to *Time* and other mass-circulation media. Matt Ridley's *Nature via Nurture* received twenty-three and Richard Lewontin's *The Triple Helix* received twenty, in a similar range of publications. Four other books, Hamer and Copeland's *Living with Our Genes,* Fausto-Sterling's *Sexing the Body,* Jonathan Pincus's *Base Instincts,* and the second edition of Susan Oyama's *Ontogeny of Information,* received fewer than ten reviews each. Each set also included reviews in trade publications, such as *Publishers' Weekly,* which give summaries and information about the publisher's marketing strategy, if any, as well as signals of the anticipated impact of the book.

Here again, numbers tell a story. Pinker's genetic and evolutionary psychology manifesto received twice the attention of Ridley's volume, in which the effects of genes are presented as tempered by environmental factors ("via nurture"). Lewontin's polemic against genetic determinism and articulation of his own version of an interactionist approach received just a bit less attention than Ridley's book. The alternatives represented by the other books—brain science (Pincus), developmental systems theory (Oyama and, to some extent, Fausto-Sterling), behavioral genetics (Hamer and Copeland)—receive much less attention. Given only the facts of book publication, one might think the various approaches received equal time. But taking into account their differential uptake shows again the power of the genetic approach, especially when engagingly presented, to command attention. In addition, all through the responses to these volumes, the question of understanding behavior continues to be represented as a matter of understanding either the etiology of individual behavior or the basis of individual variation (do our genes or our physiology make us do it? our upbringing and social environment or both?). Alternative ways of posing the question, alternative questions about behavior, especially questions about dif-

----

27. One exception to this pattern is philosopher Simon Blackburn's review of Pinker's *The Blank Slate* in the *New Republic,* a review that stressed conceptual and argumentative shortcomings. Blackburn also published a review in *New Scientist.*

ferences in frequency and distribution of behaviors in different populations, are rarely, if ever, raised. Aggression is represented as a phenomenon to be mitigated, diminished, by addressing the causes under discussion, as they "affect" individuals. Because the nature-nurture issue is so salient as the dominant controversy, even in those few cases where a population-level causal factor is mentioned, it is only considered in a context focused on factors that act on individuals.

Implications

What does it matter what kind of research gets taken up where and by whom? The empirical work analyzed in the preceding chapters concerns behaviors, some forms of which are matters of considerable public concern. Both aggression and homosexuality are valenced as social problems. Aggression research is understood as a key to understanding negatively valued behavior, from classroom disruptiveness to schoolyard bullying to violent crime. Research on sexual orientation is almost exclusively research on homosexuality, which, as recent United States politics demonstrates, still inspires fear and loathing in a significant portion of the population. Scientific research can be a corrective to misinformation and prejudice, but it is also subject to interpretation and selective reading as it moves from laboratory and field to policy deliberation, mass media, and public hearts and minds. The previous section showed how the variety of behavioral research gets reshaped for public consumption. Some areas are barely represented, in contrast with others that are treated as the source of eventual scientific understanding.

Both of the kinds of theoretical issue analyzed in previous chapters play roles in policy and in public understanding. Questions of causation are obviously relevant in connection with prevention, intervention, and attribution of responsibility. Questions of concept formation concern what is alleged to be prevented or controlled. These issues, of course, interact. Thus, when aggression or sexual orientation are understood as individual characteristics or dispositions, prevention efforts will center around methods of manipulating the causes of these dispositions as they operate on individuals. Many, if not most, geneticists profess that the complexity of the human genome argues against the prospect of finding a single mutation, or even a small number of mutations, that could be used diagnostically or as targets of manipulation. Nevertheless, opponents of genetic approaches, instigated perhaps by the less circumspect among the genetics advocates, invoke Brave New World scenarios, in which genetic engineers fix one's destiny even before birth. Less dramatic, but no less problematic, scenarios are described in which protein synthesis or activity might be pharmacologically altered to produce desired personality types.[28] More realistically in terms of available technology, some worry about the use of selective abortion on the basis of fetal genotype profiling.

Physiological research either complements genetic research or identifies primarily brain or neurological conditions responsible for or contributory to individual behavioral patterns. Much of the research on physiological determinants of aggressive behavior is read as suggesting that these too might be pharmacologically addressed. Some physiological research lends itself to treating problematic behavior as expressive of disease and thus as a medical rather than a legal (or moral) matter.[29] Because the physiological conditions associated with aggression are so various and so variously caused (genetic, congenital, brain injury, infection), this work does not so easily lend itself to hopes of systematic prevention or reduction of violence. This work might, however, be used to support claims of diminished capacity on the part of individuals accused of crime. While both hormonal and structural causal factors related to sexual orientation have been investigated, no studies have been decisive enough to encourage hopes of systematic control. Indeed, while hormonal treatments are employed to assist the development of physical secondary sex characteristics for transgender or transsexual persons, they do not seem to have an effect on sexual orientation, thus reinforcing the distinction between gender identity and sexual orientation. Physiological research on sexual orientation could of course fill in mechanisms of genetic influence, and for this reason it is regarded with suspicion by critics and opponents of genetic research.

Social environment–oriented psychological research is sometimes carried out as clinical research on possible forms of social intervention. Work that studies the effects of teaching alternative communication and interaction styles, both within families and among peers, for example, bypasses research into causes of problematic behavior and investigates the effects of intervention no matter what the cause. This work holds the promise, where sufficiently successful, of general

---

28. The widespread use of Ritalin to control ADHD provides one model of pharmacological intervention. Other psychopharmaceuticals are also in wide use despite uncertainties concerning their long-term effects. The worry of the critics is about preemptive administration of pharmacological intervention. Some thinkers also worry about the pathologization of states and behaviors within the normal human spectrum. See Angell 2011.

29. The practice of chemically castrating sex offenders can also be read as a form of medicalizing sexual offending, but the treatment of children classified as ADHD may be the most prevalent instance of medicalizing behavioral problems. See Hawthorne 2010. Many articles in general-circulation media hold out the hope for pharmacological "solutions," once the cause of a condition/behavior has been identified. See also Robinson 2009.

adoption and implementation in clinics and schools to effect changes in individuals' behavior. Other research investigates, using databases, the effects of childhood experiences such as physical abuse on later behavior. As noted earlier, some (friendly) critics of this research suggest that knowledge of biological causal factors might help in directing psychosocial forms of intervention to those most likely to be affected.[30] Less friendly critics of psychological research focused on social environment invoke the memory of B. F. Skinner's behaviorist proposals or Orwell's *1984*, and the social (re-)programming efforts of the Soviet Union. There is also a suggestion that advocates of environmentally directed intervention strategies are demanding that entire social systems be changed, with the likelihood of serious unanticipated effects. Just as the Brave New World scenario seems based on a grossly oversimplified conception of genetic processes, these criticisms of social/psychological research are based on an oversimplified view of what kinds of interventions it might support. It is hardly providing knowledge that could be used to implement such social nightmares. Indeed, the teaching of alternative communication and interaction styles is based on a quite different psychological theory than old-school behaviorism.

Finally, developmental systems theory's emphasis on the interaction of genetic, physiological, and environmental factors presents a picture of organisms sensitive to multiple influences but resistant to systematic interventions to desired outcomes. Nevertheless, contrary to DST's basic tenets, understanding relations and interactions of key factors and their causal pathways, if these are identifiable, would presumably offer insights into prospects and possible mechanisms for reducing unwanted and enhancing desired behaviors. Given the analyses above, the prospects of finding such pathways is greater for disorders than for behavior within the "normal range of variation."[31]

Whether genetic, physiological, or social-environmental, all of these approaches fuel anticipation of intervention at the level of the individual. And all involve more complicated, fine-tuned, or sophisticated versions of manipulations already practiced. Breeding, selective embryogenesis and gestation, genetic engineering, surgery, medication, and teaching and influencing through reward and punishment regimes are relatively familiar practices (at least as applied to plants and animals, if not yet all to humans), and they are geared to the production of individual organisms possessing or lacking particular traits or dispositions.

---

30. This attitude is given credibility by the Caspi and Moffitt research demonstrating the interaction of a mutation in the MAOA gene and childhood abuse.

31. As Angell (2011) notes, this distinction is not fixed.

Population approaches lend themselves to thinking, instead, in terms of societies and the differences among them: internally violent versus peaceful, externally aggressive versus neighborly, productive versus contemplative (or subsistence), monomorphic versus polymorphic as regards sexual interactions, rigidly versus flexibly monomorphic, and so on. Furthermore, the variations in frequency, intensity, diversity, or kind of interaction or behavior are related to variations in population conditions: whether structural (e.g., differences in age structure, division of labor, geographical dispersedness, or competitiveness versus cooperativeness) or resource-related (scarcity versus plenty, increasing versus diminishing, steady versus variable, homogeneous versus heterogeneous). And this would necessarily inflect the direction of intervention. Rather than bearing down on an individual (or family) within a society, it would move out from a subgroup in a society to the conditions of the entire society/population, many of which, like age structure, are themselves of such complex origin, and sensitive to so many other conditions, that they seem much less responsive to attempts at intervention. Instead, they often seem like the givens of nature. But while age structure, or even poverty and income inequality, may seem resistant to modification, more finely grained aspects of structure, such as distribution of policing or the prevalence of practices of "looking after," may well be amenable to intervention.[32]

In his 2002 book *Science, Truth and Democracy*, Philip Kitcher takes up the question of when it might be morally, if not pragmatically, legitimate to prohibit research on certain questions.[33] In the end, he argues that prohibiting research will as likely as not have the contrary effect to the one hoped for. He makes an important observation in the course of considering what kind of argument could support prohibition. Not only are the evidential status of the science and the nature of its conclusions relevant, but also the attitudes of the society in which the research will be diffused and which will affect its interpretation. What are the contexts of reception for research on aggression and on sexual orientation?

CRIME AND AGGRESSION

As previously noted, the orientation of aggression research toward crime and violence within an individualist framework means that aggression, when valenced negatively, is understood as an individual failure, whether this is under-

---

32. For example, Fagan, West, and Holland 2003.
33. Kitcher 2002.

stood morally, legally, or psychomedically. This stands in contrast with the ways in which aggression was understood when the research was oriented toward dominance (as a component of individual success), or when it is understood as part of a functional behavioral repertoire elicited by specific stimuli. Taking the unit of investigation to be the individual and his or her local group also has the consequence that crime and violence are to be understood through understanding the causal factors that dispose some individuals toward violence and criminality while other are not so disposed. The issue is perceived, that is, as investigating what makes a criminal a criminal or a delinquent a delinquent. The social problem of crime is thus understood as a problem about individuals. Even when systemic or structural factors such as distribution of income are investigated, the questions concern the impact of these factors on individual behavioral dispositions rather than on population-level behavioral properties.[34] The focus of prevention and correction is directed toward the individual, whose relations with the environment are understood as her or his behavior, which itself becomes, however acquired, an inherent characteristic of that individual. Preventing and correcting crime then become questions of either locking up the criminal or, in the best case, redirecting the potential criminal.

When no or little distinction is made among kinds of crime (excepting "white collar crime"), all convicts are treated as if they were violent and aggressive. They become the object of fear and suspicion and are treated as legitimate targets of abusive treatment by corrections officers and other inmates.[35] Since the 1970s the rate of arrests resulting in eventual incarceration has increased from 13% to 28% in the United States. Prison populations have grown as a result both of increased incarceration and longer sentences with reduced access to parole. And prisons in the United States are notorious for overcrowding and inmate vio-

---

34. A particularly dramatic example of this is provided by recent research reported in Arehart-Treichel 2009, in which the MAOA gene (the short allele of which is implicated in the Caspi and Moffitt work) was linked to gang membership and to the use of guns by gang members. Even though the researcher whose work was being discussed is quoted as saying that the research might indicate that such effects could be mitigated by environmental effects, the article frames the work as showing the effects of a genetic configuration. See also Robinson 2009, in which intervention efforts are described as targeted at "criminogenic" factors that impact individuals.

35. A quick search of Google Scholar (search term: prison guard violence) reveals few studies of prison guard brutality toward their charges in comparison with studies of inmate-on-inmate violence, and a relatively low rate of uptake measured by citation numbers. As Jacobs 2004 demonstrates, this is not because such brutality does not occur. Rather, the difficulty of filing complaints and the very nature of sequestration prevents this information from circulating. And, I am suggesting, our conception of those convicted of crimes restricts our sympathy toward them.

lence. David Garland has documented a shift of emphasis in incarceration from rehabilitation to punishment from the 1970s to the 2000s.[36] While this shift undoubtedly has complex political and economic causes, a conception of human behavior that might counteract the ideology of individual criminality that supports retributivist rather than rehabilitationist approaches is absent or unheard in the debates. Resistance to the early release of nonviolent inmates to alleviate the overcrowding in California's prisons, claiming that it would constitute a danger to society, shows the conflation of convict with violent criminal. But the largest single category of incarcerated individuals in the United States consists of drug offenders, not assailants.

It is too much of a stretch to blame the abuse of prisoners on subtle tendencies in behavioral research. Rather, I want to suggest that two pairs of narratives about crime feed into research and its public uptake. One pair employs a contrast between malevolence and illness. According to one of these narratives, the commission of particularly heinous acts is attributable to some sort of internal malfunction, an expression of some pathological state: "X must be crazy to do such a thing." According to the other, the criminal is evil. These apparently opposed narratives share a common presupposition: that behavior is to be understood in terms of causes internal to the individual. Action deemed evil has evil and malevolence in the individual as its source, and the individual has no right to expect any better than she or he meted out to her or his victims. Action deemed pathological has pathology in the individual as its source, and the individual should be cured where possible and removed from society where cure is not possible. Treating the sick/evil or nature/nurture dichotomies as the primary poles in debates about criminal behavior removes from view other research approaches that challenge the individualist presupposition. The second pair of narratives consists of two sets of narrowings. The use of indices such as arrest for violent crime in research on aggression identifies aggression with criminal violence; the effort to understand crime through research on aggression identifies crime as aggression. But there are many kinds of criminal harm (from conscious release of industrial pollutants to financial malfeasance) that do not qualify as aggression, and many occasions of incarceration that have involved no violence (for example, possession of small quantities of illegal drugs). These two narrowings support the conception of crime as violence and of the incarcerated individual as a perpetrator of violence. These conflations combined with the malevolence/illness dichotomy bode ill for the humane treatment of prisoners.

---

36. Garland 2001.

SEX AND SEXUAL ORIENTATION

The focus on individual behavior, on what factors contribute to individual preferences regarding gender of sexual partner, as well as the convenience of bimodal categories, similarly interacts with social preoccupations and assumptions concerning sex to create a type, the homosexual. This focus on the person directs both positive and negative attitudes. Like the criminal, or potential criminal, the homosexual, or potential homosexual, whatever the etiology of individual dispositions, can be cured or behaviorally redirected. The etiology determines what kinds of intervention strategy will be effective or proposed as effective.

The social and cultural situation of sexual orientation has changed in the opposite direction from that of aggression in the past twenty to thirty years. In the industrial and postindustrial world, a gay rights movement has brought homosexual relations out of the shadows. Since the removal of homosexuality from the list of mental disorders, the logic of equality and civil rights has led to legal improvements ranging from the decriminalization of sexual behavior to the lifting of bans on military service to legal changes permitting gay couples to marry. These changes are not uniform and continue to arouse strong opposition, especially in the United States and in developing countries.[37] This context has paradoxically reinforced the treatment of sexual orientation in an individualist frame, as the focus switches from sexual behavior to the homosexual as a person whose civil rights are to be protected.

Some gay rights activists have embraced the biological approaches as demonstrating that homosexuality is not a choice, and hence not a valid ground for condemnation or discrimination. Dean Hamer, who led the team investigating XQ28, continues to work (on the side) on the genetic basis of homosexual orientation in men and to engage in community advocacy and education. His efforts are directed toward promoting the genetic view as proof that same-sex erotic orientation is a matter of one's inborn constitution and not a matter of choice. Such an argument presupposes a highly dichotomized picture both of the phenomena and of the causal landscape and involves a variety of questionable assumptions about (the relations) between biological inclination, social and experiential learning, and intentionality. Since, on the genetic approach, some significant

---

37. See, for example, controversies about gay clergy in the Anglican and other churches, about gay marriage, about adoption or foster-parenting by gay couples, and about "conversion therapy." Homosexuality is still prosecutable and punishable with life sentences or death in between twenty and thirty countries.

proportion of the etiology of male homosexuality remains unexplained, one wonders if that portion could be legitimately treated as a matter of choice, and thus, as legitimately subject to sanction. And if female homosexuality is, as some researchers suggest, much more a matter of conscious reflection,[38] one wonders if that too could be legitimately subject to sanction. Apart from these problems, there's little reason to think that showing that homosexuality has a biological cause would induce many people to abandon their campaigns to remarginalize homosexuality. Indeed a genetic origin would make homosexuality much easier to control than if it were inculcated through social interaction, through cultural learning, or through some complex interaction of social, biological, and experiential factors.[39]

•

In the case of research on aggression, when the United States has an incarceration rate higher than that for any other industrialized nation, when one segment of the population is incarcerated at a much higher rate than other segments, when the sentencing practices of the world's seventh largest economy (the state of California) result in the costs of incarceration outweighing the costs of other social functions, when prisons are sites for the transformation of petty thieves into hostile, alienated, and deeply antisocial individuals, it would make sense to investigate the conditions that differentiate the United States from other industrial societies, or local variations within the United States, in terms of crime rates and incarceration. In the case of sexual orientation, replacing the binary hetero-/homosexual with more finely grained varieties of sexual/erotic interaction and understanding patterns of their variation across different social and cultural formations might reduce the discomfort aroused by sexual difference. Maybe such orientations represent phenomena that can be changed; maybe they are not in human power to change. Perhaps they are a mixture. Fixation on questions like "Is X a result of nature or of nurture?" or "What combination of nature and nurture results in X?" obscures other questions about X. Focusing on causes of individual traits or differences distracts from problems with the very conceptualization of the phenomena whose causes we seek to understand.

---

38. Peplau 2001; L. Diamond 2008.
39. See Greenberg and Bailey 2001; West 2001.

Conclusion

This review of research uptake yields several observations that ought to concern us. First, apart from isolated confrontations staged by journal editors or conference organizers, there is little interaction among proponents of different research approaches. Thus, the vaunted self-correcting processes of scientific inquiry may operate within, but not across, approaches. In part this is consistent with the incommensurabilities discussed in chapter 8. To the extent that approaches investigate relations in different causal spaces, they lack a common empirical base to ground their discussions. On the other hand, there is a common phenomenon—human behavior—or particular common phenomena—specific human behaviors, such as aggression or sexual behavior—that they purport to illuminate. The facts of empirical success in the different approaches, however partial, ought to be mutually relevant.

Second, the research itself is unevenly reported in the popular media. In both quantity and interpretation, genetic research is represented as the major and most productive line of investigation. In part, this is attributable to a national focus on genetic research in general. The Human Genome Project, for example, promised to usher in a new era not just of genomic medicine, but of human self-understanding.[40] But despite the great strides made in understanding the complexities of the genome since the structure of DNA was announced in 1953, the actual gains made in the genetics of human behaviors are most promising in illuminating behavioral or psychological disorders. And even here the results are not just partial but marginal.[41] When the details of studies—both the behavioral indices used and the meaning of the statistics reported—are not clarified in media reporting, the overall impression for readers is that genetic research is both the most successful scientific approach to understanding human behavior and the most probable source of future progress. Even though neurobiological approaches are accorded almost as much credibility as genetic approaches, the latter continue to dominate public attention.

Third, the framing, whether implicit or explicit, of behavioral research in terms of the nature-nurture debate suggests that the main question about behavior concerns individual behavior and individual differences in behavior. While the approaches focused on properties of individuals or differences among them do have a bearing on this issue, and while it is, of course, individuals who behave,

---

40. See, for example, essays in the Kevles and Hood (1993) collection.
41. See McClellan and King 2010 for a recent assessment.

it is misleading to think that all scientific questions about behavior are questions about individual behavior, or that research on individual behaviors and individual differences is adequate to address all of the issues that most concern us. While person-by-person interventions, whether in the development of aggressive or prosocial tendencies or the development of erotic preference, may alter the futures of some individuals, changing the scale of analysis may, in the one case, offer more successful strategies of intervention, and in the other, promote acceptance of different sexual patterns as consequences of large-scale social and economic transformation.

11

A Brief Conclusion

Three major points emerge from this study. One concerns the interrelations of the various approaches to the study of behavior. Another concerns the conceptualization of the phenomena to be explained. A third concerns the communication of research findings within the research community and in the wider public arena.

Each of the approaches, I have argued, is partial, equipped by its investigative tools to explore particular portions of the causal space but not the whole of it. Given that each represents the causal space differently, the sense in which these approaches can be said to offer knowledge must be explained.[1] One might be tempted to think that each approach can fully explain some portion of the etiology or distribution of a given behavior in a population. Put them all together and eventually one would have a complete causal story. But this could only be the case if the approaches parsed the causal space in the same way, that is, if each recognized all possible causal factors in the space, put the same contents in each category, and were able to separate the effects of one type from the effects of the other types as well as to represent their interactions. I have argued, however, that even if it were possible to construct models that included all of the putative factors and their possible interactions and that represented their causal spaces identically, it would not be possible to obtain the simultaneous measurements

---

1. I am grateful to Greg Priest for urging me to clarify this point.

of every value in any actual situation. And without such measurement, it would not be possible to obtain empirical confirmation of an additive model of the factors' direct and interactive effects. In practice, measuring the value of any one parameter requires holding the others constant. When those other parameters are themselves of different kinds, the contribution they may be making to the phenomenon or to the effect attributed to the factor being measured cannot be identified. The kind of knowledge afforded by the approaches is not like pieces of a puzzle that combine to make a complete and coherent picture. Instead, knowledge claims must be conditioned on particular parsings of the causal space. An approach can indicate certain causal dependencies assuming the values of other causal factors remain constant. But this is just what fails to hold in the actual world. This failure, which we could more positively term the world's complexity and dynamism, is what the experiments of the developmental systems researchers show in the case of behavior. They have not, however, succeeded in offering either hypotheses that fully incorporate all the complexity they claim characterizes organic processes or a strategy for empirically demonstrating such a hypothesis.

Experiments or studies performed from within any given approach tend to show that some fraction of the phenomenon is associated with or attributable to a causal factor investigated by that approach. With few exceptions, any particular factor or type of factor correlates with at best 50% of the variance in expression of a behavioral trait in a human population, and often raises the probability of expression only slightly over the base rate. The benefit of finding any such association is that it constitutes a point of entry for additional investigation. Holding other factors constant (or assuming them to be so), it is possible to explore the influences, interrelations, and interdependencies of factors in one's chosen area of investigation. Thus our knowledge of the organism is increased, even though our understanding of the supposed target phenomenon may not be. In that sense, the more we know, the less we know. In fact, one might even characterize the knowledge produced as counterfactual in nature: were all else inactive or held constant, a change in the value of X by n in direction T would be accompanied or followed by a change in the value of Y by m in direction S.[2]

Where interactions among factors belonging to different regions of the causal space are pursued, the greatest success has accrued to approaches studying spe-

---

2. The counterfactual is similar to *ceteris paribus* laws, but the content of knowledge in these cases is not otherwise lawlike.

cific psychiatric disorders with specific neural bases. Yet even here, we are given probabilistic relations. Some significant proportion of the individuals manifesting the condition or expressing the behavior in question do not or will not fit the causal picture, whatever causal picture is in question. And the population approach that, by definition, covers all cases will not reveal anything about what differentiates one individual from another.[3]

Neither the popular nor the professional versions of the nature-nurture debate accurately represent this epistemological situation. Furthermore, the question that prompts the debate in either form is based on a false presupposition. The answer to the popular question "Are we made by our genes or by our environment?" is not just "By both" but also "It's incorrect to suppose that it must be one and/or the other." And the answer to the professional question "Which approach is going to yield the most information about the etiology of behavior?" is not just "All of them, depending on what one wants to do with the knowledge," but also "It's incorrect to suppose any single approach could provide a more comprehensive picture than any other." Once it is acknowledged that multiple causal factors belonging to different theoretical/ontological domains are involved in the inculcation of any given behavioral disposition, the idea that some single approach will yield a complete and precise account that covers all or most of the factors becomes a map for a fool's errand. The view of the developmental systems theorists, that all factors interact with each other, is probably correct, but, as I have said in earlier chapters, in a metaphysical rather than an empirical sense. Empirically, the research that can be called on to support that view is itself performed under the aegis of more restrictive frameworks.[4] The experimental work involves contrasting and combining the effects of not more than two factors (usually presumed genetic and some specific feature of the environment), not the full range posited by the theory. Moreover, empirically successful integrationist approaches, such as that of Caspi and Moffitt, achieve their success by restricting the explanandum to acknowledged forms of disorder or mental illness, thus leaving a larger range of behavior unexplained. And, as previously stressed, they do not account for more than a fraction of the incidence of the disorders they address. In this light, debates among proponents of the ap-

---

3. See Ostrom 2007 for discussion of this point.

4. I don't want to suggest that DST or something like it will never achieve empirical grounding (see van Fraassen 2009 for discussion of this concept) but that current empirical procedures are not adequate to do so.

proaches begin to look less like substantive questions, to which there might be a principled solution, than a dimension of the jockeying for uptake and funding that is part of contemporary science.

The scientific knowledge of human behavior that results from these multiple types of investigation will remain piecemeal, kaleidoscopic. The completeness we see through a kaleidoscope is an illusion achieved through mirrors and the obscuring of alternative configurations. There are actual relations among the elements, but the structure of the kaleidoscope prevents their being fully perceivable, suggesting only, as we turn the instrument, that there are multiple ways of perceiving them. Understanding the image produced requires appreciating its partiality. Genuine understanding of human behavior requires not a new comprehensive paradigm so much as an understanding of the scope and limitations of the various approaches employed in its investigation.

The second point is more problematic. There is no standard way of identifying and distinguishing behaviors one from another. The concept of behavior itself does not have a fixed meaning either among behavioral researchers or among philosophers, and the specific behaviors that are studied empirically have either multiple (as in the case of aggression) or contested (as in the case of sexual orientation) operationalizations. The habit of listing behaviors or dispositions in which there is a social interest, such as aggression or sexual orientation, along with conditions that can be assigned a genetic identifier, like cystic fibrosis or Huntington's chorea, feeds the twin illusions that those behavioral dispositions (1) have an unequivocal phenotypic description and (2) can be given a causal account similar in structure to the account for those undeniably genetic diseases. Neither of these notions is borne out under examination. With respect to the first, aggression is multiply operationalized, the operationalizations are differently assessed, and there is as yet no conclusive demonstration that there is an actual coherent phenomenon that can be investigated rather than a range of behavioral responses that may be elicited to different degrees and in different ways in different circumstances. In contrast, research on sexual orientation is criticized for being too narrow in its choice of objects and in its operationalization of the phenomenon to be explained. One dimension of variation is chosen for investigation, creating the impression that it is the only one, and that dimension is treated as a binary, obscuring gradations and values other than those at the two poles. When a type of behavior is operationalized for purposes of study, the hope is that a discrete phenomenon is being identified, but the variety of operationalizations and the instability across measurement methods gives the lie to such hopes. With respect to the second illusion, the most successful identifications *are* of disorders that have some organic basis. For these, the supposition that a causal account similar in structure to those given for organic conditions like Huntington's or hemophilia is not that fanciful. But it is not clear that all instances of a behavior phenomenologically identified will turn out to have an organic basis. The Caspi and Moffitt interactionist approach, for example, still leaves a significant proportion of incidences of the behaviors studied outside the range of those correlated with given gene-environment intersections. In communication to a broader public, the interpretation of research on nondisordered behavior is parasitic on the success of research on clearly defined organic disorders. Thus, by metonymy, behavior becomes pathology.

The third point, concerning the internal and external social dimensions of science, interacts with the issues of conceptualization just summarized. The analysis of citation patterns suggests that there is little interaction across research approaches and that most citations are to other work within the citing article's approach. Thus, critical interaction that could prompt a reevaluation of an approach is rare and overwhelmed by reception that essentially confirms the approach, if not its implementation in any given study. Peer review is limited in its effectiveness to intra-approach reflection and pruning. While knowledge indeed results from this research, less attention is paid to just what phenomena it is knowledge of than to the putative revelation of the causal import of the type of factor studied. Behavior may sometimes best be viewed as a point of entry into a complex causal pathway, helpful in revealing the influence of a gene, hormone, or form of social experience on a more proximate effect such as the availability of a protein, a physiological reaction, or an attitude. Nevertheless, because behavior is what the ultimate audience for the research is interested in, it ends up as the purported target of explanation in communications outside of the immediate research context.

Political critics of behavioral research, such as those described in chapter 1, tend to focus on genetic research and to decry biological determinism. But any causal inquiry, social, psychological, or biological, will have deterministic overtones. Whether one takes the side of nature or of nurture in the nature-nurture debate, one is equally committed (or noncommitted) to a deterministic picture. Of greater concern is the interaction between the conceptualization of behavior in scientific research and our social concerns. However it is operationalized in human behavioral research, aggression is represented as unauthorized infliction of harm on another (whether through physical contact or verbal abuse). This is aggression as it is represented in crime stories in the evening news. It is what we fear when walking in unfamiliar (or sometimes familiar) neighborhoods. The indices actually measured are generally surrogates for the behavior that is purportedly being studied. There are questions about the representativeness of these surrogates, as well as about the degree of coincidence of the different forms of their measurement. Self-report does not coincide with third-person report, and the passage of time changes self-reports; even criminal convictions may more accurately represent susceptibility to arrest than inclination to violence. Furthermore, the forms of aggression of interest represent a narrow portion of the spectrum of aggressive behavior, and overt aggression itself represents only a portion of behavior that could be called antisocial. But individual infliction of harm is something we wish and believe can be isolated and controlled. Thus, it is what we look to research to inform us about. Scientific research on aggression, while it may result in increased knowledge about aspects of human functioning, may be both too fragmented and too narrowly focused to satisfy this need.

The restricted understanding of sexual orientation similarly reflects social concerns. While erotic preference is not limited in its scope to sex or gender of sexual partner, the latter is the object of study. To some extent this is the legacy of nineteenth-century concerns with demography and, hence, with social reproduction. Those concerns have been largely replaced in the industrial world, by cultural preoccupations with gender relations as an avatar of social change overall.[5] Resistance to acceptance, or even tolerance, of homosexual relations tends to be associated with a preference for traditional gender roles and discomfort with changing roles for women, especially with women's greater sexual self-determination. That both social attitudes are often justified by appeals to religious teachings does not negate the fact of their co-occurrence. It is hard to escape the conclusion that the focus on homo- and heterosexuality obscures the range of erotic phenomena that constitute human sexuality.

The issues of scale raised by the availability of population-level approaches suggest that the nature-nurture debate is itself the problem. Consider research on aggression and its presumed relevance to crime. The nature-nurture lens facilitates conceptualizing the commission of crime as a result of pathological genes or pathological parenting. Combined with racial disparities in incarceration and a national history of racial bias, this conceptualization facilitates a further association of African ancestry with one or the other of those pathologies. One need only remember the proposals that differences in average IQ scores among racial groups were the outcome of heritable differences (generally understood to be

---

5. Much of the opposition to gay marriage, for example, has come from religious groups and political organizations that also promote traditional forms of the sexual division of labor.

genetic), or the handwringing over the rate of single-parent households among urban African Americans, to remember how available such explanations for differential social outcomes are.[6] By failing to incorporate population perspectives into their thought or reporting, perspectives that might suggest quite different questions than those focused on etiologies of individual difference, debaters about nature versus nurture are unwittingly complicit with a system of racial subjugation that seeks legitimacy in science. Similarly, looking for the "causes" of homosexuality in genes or hormones, on the one hand, or problematic relations with parents, on the other, in isolation from similar questions about heterosexuality, perpetuates the notion that homosexuality is a problem rather than one dimension of variation that finds different degrees of expression in differently structured social worlds.

•

Whether an operationalized behavior is in the family of aggressive behaviors or of sexual behaviors, it can be studied as a characteristic of individuals or as a characteristic of populations. When behavior is understood individualistically, not only is it treated as the action of an individual but its etiology is presumed to involve factors internal to the individual, whether these be genes, neural circuitry, or psychological orientations resulting from internal or external influences. Given the moral interest in individual behavior, our responses to expressions of aggressive or sexual dispositions are inflected with attitudes of praise or blame or exculpation. When understood as a population characteristic, the focus shifts from individual dispositions to the frequency and distribution of behaviors. Here moral attitudes are idle. It does not matter what the causes of individual dispositions are when patterns of frequency or distribution are the concern. Population-level causes are impersonal; they give no insight into individual motivation, even though they are structural or processual conditions to which differently situated individuals respond. As impersonal phenomena, they do not excite our moral passions in the way personal characteristics, however inculcated, do.

These moral passions, however, stand in the way of empirically informed social policies and of a more complete presentation of the range of research questions and approaches that are relevant to such policies. Of course, individuals

---

6. For a detailed account of the availability of such explanations in entertainment media and in social science, see Covington 2010.

commit crimes and individuals engage in sexual behavior. And in many cases, individuals suffer pathological conditions that can be individually addressed. If, however, we continue to seek the causes of behavior solely within the individual, we may fail to perceive, let alone understand, larger patterns in the variations of frequency and distribution. Without such understanding, we will stay mired in the attitudes of praise or blame versus exculpation, or punishment versus cure, appropriate to individuals. If the condition or behavior of an individual is the issue, then the factual questions whose answers ground these attitudes are the appropriate questions. Punishment and cure, praise and blame, are after the fact. In spite of the deterrence theory of punishment, they do little to affect the overall incidence. If public policy is to be guided by science, our first concern ought to be to insure that the full range of scientific knowledge is available. Even when research does not point unequivocally to some policy, it may help to avoid courses of action with uncertain or even harmful outcomes. Moreover, forthrightness about the limits and plurality of scientific knowledge will reduce the chances of its being discredited when so-called science-based policies fail.

I end with some questions to which I do not yet know the answers. While American individualism and methodological reductionism undoubtedly play some part, what explains the emphasis on individual-centered research and reporting of research?[7] The previous chapter's preliminary review of public uptake shows biases in favor both of reporting results from genetic research and of representing behavioral questions as concerning individual traits and variation. What accounts for these biases? Are they transferred from the research? Do they play any role in shaping the research? It is easy enough to note *how* research is presented; explaining *why* it is so represented is another matter entirely. And if the patterns of uptake in the professional literature indicate little cross-approach interaction, is an ideal of scientific objectivity being compromised? Should measures be considered that would promote such interaction? A pluralist perspective is not the end of analysis but the beginning of new questions.

---

7. See Haslanger (2011) for further discussion of the resistance to thinking above the level of individuals.

Appendix

Table 1.   Citations to a subset of articles by Caspi and Moffitt

Category	Including books		Excluding books	
Research	136	35.7%	134	35.9%
Clinical	56	14.7%	54	14.5%
Clinical/research	136	35.7%	132	35.4%
Policy	9	2.4%	9	2.4%
Policy/research	8	2.1%	8	2.1%
Policy/clinical	2	0.5%	2	0.5%
Policy/clinical/research	22	5.8%	22	5.9%
General interest	4	1.0%	4	1.1%
Miscellaneous	8	2.1%	8	2.1%
Total	**381**		**373**	

Based on Moffitt et al. 2004; Moffitt 2005a, 2005b; Moffitt, Caspi, and Rutter 2005.

Table 2.   Citations to a subset of articles by Gottlieb and collaborators

Category	Including books		Excluding books	
Research	486	79.7%	384	72.9%
Clinical	18	3.0%	18	3.7%
Clinical/research	48	7.9%	45	9.3%
Policy	5	0.8%	0	—
Policy/research	6	1.0%	1	0.2%
Policy/clinical	1	0.2%	0	—
Policy/clinical/research	4	0.7%	3	0.6%
General interest	1	0.2%	1	0.2%
Miscellaneous	41	6.7%	34	7.0%
Total	**610**		**486**	

Based on Gottlieb 1991, 1997, 2001; Gottlieb, Wahlsten, and Lickliter 2006.

Table 3.   Citations to a subset of articles by Coccaro and collaborators

Category	Including books		Excluding books	
Research	57	36.1%	48	34.5%
Clinical	20	12.7%	17	12.2%
Clinical/research	66	41.8%	60	43.2%
Policy	1	0.6%	1	0.7%
Policy/research	4	2.5%	3	2.2%
Policy/clinical	0	—	0	—
Policy/clinical/research	8	5.1%	8	5.8%
General interest	1	0.6%	1	0.7%
Miscellaneous	1	0.6%	1	0.7%
Total	**158**		**139**	

Based on Coccaro, Kavousi, and Lesser 1992; Lee and Coccaro, 2001; Best, Williams, and Coccaro 2002.

Works Cited

Adam, Barry. 1985. "Age, Structure, and Sexuality: Reflections on the Anthropological Evidence on Homosexual Relations." *Journal of Homosexuality* 11 (34): 19–33.

Albert, D. J., M. L. Walsh, and R. H. Jonik. 1993. "Aggression in Humans: What Is Its Biological Foundation?" *Neuroscience and Biobehavioral Reviews* 17:405–25.

Alexander, Michelle. 2010. *The New Jim Crow*. New York: New Press.

Alia-Klein, Nelly, Rita Z. Goldstein, Aarti Kriplani, Jean Logan, et al. 2008. "Brain Monoamine Oxidase A Activity Predicts Trait Aggression." *Journal of Neuroscience* 28 (19): 5099–5104.

American Psychiatric Association. 1987. *Diagnostic and Statistical Manual of Mental Disorders*. 3rd ed., rev. Washington, DC: American Psychiatric Association.

———. 1994. *Diagnostic and Statistical Manual of Mental Disorders*. 4th ed. Washington, DC: American Psychiatric Association.

Anderson, C. M. 1990. "Desert, Mountain and Savanna Baboons: A Comparison with Special Reference to the Suikerbosrand Population." In *Baboons, Behaviour and Ecology, Use and Care: Selected Proceedings of the 12th Congress of the International Primatological Society*, edited by M. Thiego de Mello, A. Whiten, and R. W. Byrne, 89–103. Brasilia, Brazil: Universaded de Brasilia.

Anderson, Craig A., and Brad J. Bushman. 2002. "Human Aggression." *Annual Review of Psychology* 53:27–51.

Anderson, Gail. 2007. *Biological Influences on Criminal Behavior*. Boca Raton, FL: CRC Press/ Taylor and Francis.

Anderson, John E., and Ron Stall, 2002. "Increased Reporting of Male-to-Male Sexual Activity in a National Survey." *Sexually Transmitted Diseases* 29 (11): 643–46.

Angell, Marcia. 2005. *The Truth about the Drug Companies*. Rev. ed. New York: Random House.

———. 2011. "The Epidemic of Mental Illness. Why?" *New York Review of Books* 58 (June 23): 20–22.

Archer, John. 1991. "The Influence of Testosterone on Human Aggression." *British Journal of Psychology* 82:1–28.

Arehart-Treichel, Joan. 2009. "MAOA Gene Reveals Clues about Propensity for Violence." *Psychiatric News*, August 7, 24.

Austin, Peter C., Muhammad M. Mamdani, David N. Juurlink, and Janet E. Hux. 2006. "Testing Multiple Statistical Hypotheses Resulted in Spurious Associations: A Study of Astrological Signs and Health." *Journal of Clinical Epidemiology* 59:964–69.

Bailey, J. Michael, and Richard Pillard. 1991. "A Genetic Study of Male Sexual Orientation." *Archives of General Psychiatry* 48 (12): 1089–96.

Bailey, J. Michael, Richard Pillard, Michael E. Neale, and Yvonne Agyei. 1993. "Heritable Factors Influence Sexual Orientation in Women." *Archives of General Psychiatry* 50 (3): 217–23.

Bailey, J. Michael, and Kenneth Zucker. 1995. "Childhood Sex-Typed Behavior and Sexual Orientation: A Conceptual Analysis and Quantitative Review." *Developmental Psychology* 31 (1): 43–55.

Barrett, Louise, 2009. "A Guide to Practical Babooning." *Evolutionary Anthropology* 18:91–102.

Baumrind, Diana. 1991. "Parenting Styles and Adolescent Development." In *Encyclopedia on Adolescence*, edited by Jeanne Brooks-Gunn, Richard Lerner, and Anne Peterson. New York: Garland.

———. 1993. "The Average Expectable Environment Is Not Good Enough: A Response to Scarr." *Child Development* 64:1299–1317.

Beardslee, W. R., E. M. Versage, E. J. Wright, P. Salt, P. C. Rothberg, K. Drezner, and T. R. G. Gladstone. 1997. "Examination of Preventive Interventions for Families with Depression: Evidence of Change." *Development and Psychopathology* 9 (1): 109–30.

Beauchaine, Theodore, Zvi Strassberg, Michelle Kees, and Deborah Drabick. 2002. "Cognitive Response Repertoires to Child Noncompliance by Mothers of Aggressive Boys." *Journal of Abnormal Child Psychology* 30 (1): 89–102.

Bechtel, William, and Robert Richardson. 1993. *Discovering Complexity: Decomposition and Localization as Strategies in Scientific Research*. Princeton, NJ: Princeton University Press.

Bernet, William, Cindy L. Vencak-Jones, Nita Farahany, and Stephen A. Montgomery. 2007. "Bad Nature, Bad Nurture, and Testimony Regarding MAOA and SLC6A4 Genotyping at Murder Trials." *Journal of Forensic Sciences* 52 (6): 1362–71.

Best, Mary, J. Michael Williams, and Emil F. Coccaro. 2002. "Evidence for a Dysfunctional Prefrontal Circuit in Patients with an Impulsive Aggressive Disorder." *Proceedings of the National Academy of Sciences* 99 (12): 8448–53.

Bierman, Karen, and David Smoot. 1991. "Linking Family Characteristics with Poor Peer Relations: The Mediating Role of Conduct Problems." *Journal of Abnormal Child Psychology* 19 (3): 341–56.

Bierman, Karen, David Smoot, and Kathy Aumiller. 1993. "Characteristics of Aggressive-Rejected, Aggressive (Nonrejected), and Rejected (Nonaggressive) Boys." *Child Development* 64:139–51.

Billings, Paul, Jonathan Beckwith, and Joseph Alper. 1992. "The Genetic Analysis of Human Behavior: A New Era?" *Social Science and Medicine* 35 (3): 227–38.

Blank, Hanne. 2012. *Straight: The Surprisingly Short History of Heterosexuality.* Boston: Beacon Press.

Blumstein, Alfred, 1986. *Criminal Careers and "Career Criminals."* Washington, DC: National Academy Press.

———. 1997. "Interaction of Criminological Research and Public Policy." *Journal of Quantitative Criminology* 12 (4): 349–61.

———. 1998. "U.S. Criminal Justice Conundrum: Rising Prison Populations and Stable Crime Rates." *Crime and Delinquency* 44 (1): 127–35.

Bocklandt, Sven, Steve Horvath, Eric Vilain, and Dean H. Hamer. 2006. "Extreme Skewing of X-Chromosome Inactivation in Mothers of Homosexual Men." *Human Genetics* 118 (6): 691–94.

Bocklandt, Sven, and Eric Vilain. 2007. "Sex Differences in Brain and Behavior: Hormones Versus Genes." *Advances in Genetics* 59:245–66.

Bontempo, Daniel E., and Anthony R. D'Augelli. 2002. "Effects of At-School Victimization and Sexual Orientation on Lesbian, Gay, or Bisexual Youths' Health Risk Behavior." *Journal of Adolescent Health* 30:364–74.

Bonthron, Diana, Lisa Strain, Sarah Bundey, Elizabeth Norman, Patrician N. Howard-Peebes, Anne Maddalena, Susan H. Black, and Joseph D. Schulman. 1993. "Population Screening For Fragile-X Syndrome." *Lancet* 341:769–70.

Booth, Alan, Douglas Granger, Allan Mazur, and Katie T. Kivlighan. 2006. "Testosterone and Social Behavior." *Social Forces* 85 (1): 167–91.

Booth, B. E., M. Verma, and R. S. Beri. 1994. "Fetal Sex Determination in Infants in Punjab, India: Correlations and Implications." *British Medical Journal* 309:1259–61.

Bouchard, Thomas J., Jr., and Matt McGue. 2003. "Genetic and Environmental Influences on Human Psychological Differences." *Journal of Neurobiology* 54 (1): 4–45.

Brannigan, Augustine, William Gemmell, David Pevalin, and Terrance Wade. 2002. "Self-Control and Social Control in Childhood Misconduct and Aggression: The Role of Family Structure, Hyperactivity, and Hostile Parenting." *Canadian Journal of Criminology* 44 (2): 119–42.

Brondizio, Eduardo, Elinor Ostrom, and Oran Young. 2009. "Connectivity and the Governance of Multilevel Social-Ecological Systems: The Role of Social Capital." *Annual Review of Environment and Resources* 34:253–78.

Brown, Windy M., Christopher H. Finn, Bradley Cooke, and S. Marc Breedlove. 2002. "Differences in Finger-Length Ratios between Self-Identified 'Butch' and 'Femme' Lesbians." *Archives of Sexual Behavior* 31 (1): 123–28.

Brunner, H. G. 1995. "MAOA Deficiency and Abnormal Behavior: Perspectives on an Association." In *Genetics of Criminal and Antisocial Behavior* (Ciba Foundation Symposium 194), 155–164. Chichester: Wiley.

Brunner, H. G., M. Nelen, X. O. Breakefield, H. H. Ropers, and B. A. van Oost. 1993. "Abnormal Behavior Associated with a Point Mutation in the Structural Gene for Monoamine Oxidase A." *Science* 262:578–80.

Brunner, H. G., M. Nelen, P. van Zandvoort, N. G. Abeling, A. H. van Gennip, E. C. Wolters, M. A. Kuiper, H. H. Ropers, and B. A. van Oost. 1993. "X-Linked Borderline Mental Retardation with Prominent Behavioral Disturbance: Phenotype, Genetic Localization, and Evidence for Disturbed Monoamine Metabolism." *American Journal of Human Genetics* 52:1032–39.

Burgess, Robert C., and Peter Molenaar. 1993. "Human Behavioral Biology." *Human Development* 36:36–54.

———. 1995. "'Some Conceptual Deficiencies in "Developmental" Behavior Genetics': Comment." *Human Development* 38:159–64.

Buss, A. H., and M. Perry. 1992. "The Aggression Questionnaire." *Journal of Personality and Social Psychology* 63 (3): 452–59.

Butte, Atul. 2008. "The Ultimate Model Organism." *Science* 320:325–27.

Butte, Atul, and Isaac Kohane. 2006. "Creation and Implications of a Phenome-Genome Network." *Nature Biotechnology* 24 (1): 55–62.

Byne, William, and Bruce Parsons. 1993. "Human Sexual Orientation: The Biologic Theories Reappraised." *Archives of General Psychiatry* 50 (3): 228–39.

Byne, William, Stuart Tobet, Linda Mattiace, Mitchell Lasco, Eileen Kemether, Mark A. Edgar, Susan Morgello, Monte S. Buchsbaum, and Liesl B. Jones. 2001. "The Interstitial Nuclei of the Human Anterior Hypothalamus: An Investigation of Variation with Sex, Sexual Orientation, and HIV Status." *Hormones and Behavior* 40:86–92.

Cairns, Robert. 1991. "Multiple Metaphors for a Singular Idea." *Developmental Psychology* 27 (1): 23–26.

Cantor, James M., Ray Blanchard, Andrew D. Paterson, and Anthony F. Bogaert. 2002. "How Many Gay Men Owe Their Sexual Orientation to Fraternal Birth Order?" *Archives of Sexual Behavior* 31 (1): 63–71.

Cartwright, Nancy. 1983. *How the Laws of Physics Lie.* Oxford: Oxford University Press.

———. 2006. "Well-Ordered Science: Evidence for Use." *Philosophy of Science* 73:981–90.

Caspi, Avshalom, J. McClay, T. Moffitt, J. Mill, J. Martin, I. W. Craig, A. Taylor, and R. Poulton. 2002. "Role of Genotype in the Cycle of Violence in Maltreated Children." *Science* 297:851–54.

Caspi, Avshalom, and Terrie Moffitt. 2006. "Gene-environment Interactions in Psychiatry: Joining Forces with Neuroscience." *Nature Reviews: Neuroscience.* 7:583–90.

Caspi, Avshalom, Terrie Moffitt, M. Cannon, J. McClay, R .Murray, H. Harrington, A. Taylor, L. Arseneault, B. Williams, A. Braithwaite, R. Poulton, and I. W. Craig. 2005. "Moderation of the Effect of Adolescent-Onset Cannabis Use on Adult Psychosis by a Functional Polymorphism in the Catechol-O-Methyltransferase Gene: Longitudinal Evidence of a Gene x Environment Interaction." *Biological Psychiatry* 57 (10): 1117–27.

Caspi, Avshalom, Karen Sugden, Terrie Moffitt, Alan Taylor, Ian W. Craig, Nona Lee Harrington, Jospeh McClay, Jonathan Mill, Judy Martin, Antony Braithwaite, and Richie Poulton. 2003."Influence of Life Stress on Depression: Moderation by a Polymorphism in the 5-HTT Gene." *Science* 301:386–89.

Catton, William. 1994. "Foundations of Human Ecology." *Sociological Perspectives* 37 (1): 75–95.

Chan, Cecilia Lai-wan, Eric Blyth, and Celia Hoi-yan Chan. 2009. "Attitudes to and Practices Regarding Sex Selection in China" *Prenatal Diagnosis* 26 (7): 610–13

Cervone, Daniel, and Yuichi Shoda. 1999. "Introduction." In *The Coherence of Personality: Social-Cognitive Bases of Consistency, Variability, and Organization,* edited by Daniel Cervone and Yuichi Shoda, 3–31. New York: Guilford Press.

Charles, D. 1992. "Genetics Meeting Halted amid Racism Charges." *New Scientist* 135:4.

Cherek, Don, Scott D. Lane, Cynthia J. Pietras, and Joel Steinberg. 2002. "Effects of Chronic Paroxetine Administration on Measurements of Aggressive and Impulsive Responses of Adult Males with a History of Conduct Disorder." *Psychopharmacology* 159:266–74.

Chiang, Howard Hseuh-Hao. 2009. "Homosexual Behavior in the United States, 1988–2004: Quantitative Empirical Support for the Social Construction Theory of Sexuality." *Electronic Journal of Human Sexuality* 12. http://www. ejhs.org/Volume 12/Homosexuality.htm.

Chu, Junhong. 2001. "Prenatal Sex Determination and Sex-Selective Abortion in Rural Central China." *Population and Development Review* 27 (2): 259–81.

Churchland, Patricia. 1989. *Neurophilosophy: Toward a Unified Theory of the Mind/Brain.* Cambridge, MA: MIT Press.

Clifton, Allan, Eric Turkheimer, and Thomas F. Oltmann. 2007. "Improving Assessment of Personality Disorder Traits through Social Network Analysis." *Journal of Personality* 75: 1007–32.

Coccaro, Emil. 1993. "Psychopharmacologic Studies in Patients with Personality Disorders: Review and Perspective." *Journal of Personality Disorders*, supplement (Spring), 181–92.

Coccaro, Emil, C. S. Bergeman, Richard J. Kavoussi, and A. D. Seroczynski. 1997. "Heritability of Aggression and Irritability: A Twin Study of the Buss-Durkee Aggression Scales in Adult Male Subjects." *Biological Psychiatry* 41 (3): 273–84.

Coccaro, Emil, Steven Gabriel, and Larry Siever. 1990. "Buspirone Challenge: Preliminary Evidence for a Role for Central 5-HT-1a Receptor Function in Impulsive Aggressive Behavior in Humans." *Psychopharmacology Bulletin* 26 (3): 393–405.

Coccaro, Emil, Richard J. Kavoussi, and J. C. Lesser. 1992. "Self- and Other-Directed Human Aggression: The Role of the Central Serotonergic System." *International Clinical Psychopharmacology* 6 (supp. 6): 70–83.

Coccaro, Emil, Howard Klar, and Larry Siever. 1994. "Reduced Prolactin Response to Fenfluramine Challenge in Personality Disorder Patients Is Not Due to Deficiency of Pituitary Lactotrophs." *Biological Psychiatry* 36:344–46.

Coccaro, Emil, Jeremy M. Silverman, Howard M. Klar, Thomas B. Horvath, and Larry J. Siever. 1994. "Familial Correlates of Reduced Central Serotonergic System Function in Patients with Personality Disorders." *Archive of General Psychiatry* 51:318–24.

Cohen, Kenneth. 2002. "Relationships among Childhood Sex-Atypical Behavior, Spatial Ability, Handedness, and Sexual Orientation in Men." *Archives of Sexual Behavior* 31 (1): 129–43.

Cohen, Patricia. 2011."Genetic Basis for Crime: A New Look." *New York Times*, June 20. http://www.nytimes.com (accessed June 24, 2011).

Constantino, John N. 1995. "Early Relationships and the Development of Aggression in Children." *Harvard Review of Psychiatry* 2:259–73.

Covington, Jeanette, 2010. *Crime and Racial Constructions: Cultural Misinformation about African-Americans in Media and Academia.* Lanham, MD: Lexington Books.

Crick, Francis. 1993. *The Astonishing Hypothesis: The Scientific Search for the Soul.* New York: Scribner's.

Crick, Nicki R., and Kenneth A. Dodge. 1996. "Social-Information-Processing Mechanisms in Reactive and Proactive Aggression." *Child Development* 67:993–1002.

Dabbs, James M., Timothy S. Carr, Robert L. Frady, and Jasmin K. Riad. 1995. "Testosterone, Crime, and Misbehavior among 692 Male Prison Inmates." *Personality and Individual Differences* 18 (5): 627–33.

Damasio, Antonio. 1999. *The Feeling of What Happens*. New York: Harcourt.

Del Vecchio, Tamara, and Susan O'Leary. 2006. "Antecedents of Toddler Aggression." *Journal of Clinical Child and Adolescent Psychology* 35 (2): 194–202.

Dennett, Daniel. 1991. *Consciousnesss Explained*. New York. Little Brown.

DeVries, B. B., D. J. Halley, B. A. Oostra, and N. J. Niermejeier. 1998. "The Fragile X Syndrome." *Journal of Medical Genetics* 35 (7): 579–89.

de Waal, Frans. 1989. *Peacemaking among Primates*. Cambridge, MA: Harvard University Press.

———. 1992. "Aggression as a Well-Integrated Part of Primate Social Relationships." In *Aggression and Peacefulness in Human and Other Primates*, edited by James Silverberg and J. Patrick Gray, 37–56. New York: Oxford University Press.

de Waal, Frans, and Frans Lanting. 1997. *Bonobo: The Forgotten Ape*. Berkeley: University of California Press.

Diamond, Lisa M. 2008. *Sexual Fluidity*. Cambridge, MA: Harvard University Press.

Diamond, Milton. 1993. "Homosexuality and Bisexuality in Different Populations." *Archives of Sexual Behavior* 22 (4): 291–310.

———. 2004. "Sex, Gender, and Identity over the Years: A Changing Perspective." *Child and Adolescent Psychiatric Clinics of North America: Sex and Gender* 13:591–607.

Dickemann, Mildred. 1993. "Reproductive Strategies and Gender Construction." *Journal of Homosexuality* 24 (3): 55–71.

DiLalla, Lisabeth Fisher. 2002. "Behavior Genetics of Aggression in Children: Review and Future Directions." *Developmental Review* 22 (4): 593–622.

DiLalla, Lisabeth Fisher, and Irving Gottesman. 1991. "Biological and Genetic Contributors to Violence: Widom's Untold Tale." *Psychological Bulletin* 109 (1): 125–29.

Doell, Ruth, and Helen E. Longino. 1988. "Hormones and Human Behavior: A Critique of the Linear Model." *Journal of Homosexuality* 15 (3/4): 55–78.

D'Onofrio, Brian, Wendy S. Slutske, Eric Turkheimer, Robert E. Emery, K. Paige Harden, Andrew C. Heath, Pamela A. F. Madden, Nicholas G. Martin. 2007. "Intergenerational Transmission of Childhood Conduct Problems: A Children of Twins Study." *Archives of General Psychiatry*. 64 (7): 820–29.

Dörner, Gunther. 1988. "Neuroendocrine Response to Estrogen and Brain Differentiation in Heterosexuals, Homosexuals, and Transsexuals." *Archives of Sexual Behavior* 17 (1): 57–75.

Dretske, Fred. 1988. *Explaining Behavior*. Cambridge, MA: MIT Press.

Duman, Darah, and Gayla Margolin. 2007. "Parents' Aggressive Influences and Chidren's Aggressive Problem Solutions with Peers." *Journal of Clinical Child and Adolescent Psychology* 36 (1): 42–55.

Eckert, Elke, Thomas Bouchard, J. Bohlen and L. L. Heston. 1986. "Homosexuality in Monozygotic Twins Reared Apart." *British Journal of Psychiatry* 148:421–25.

Edelman, Gerald, W. E. Gall, and W. M. Cowan, eds. 1984. *Molecular Bases of Neural Development*. New York: Wiley and Sons.

Edelman, Gerald, and Giulio Tononi. 2000. *A Universe of Consciousness*. New York: Basic Books.

Ehrenkranz, Joel, Eugene Bliss, and Michael H. Sheard. 1974. "Plasma Testosterone: Correlation with Aggressive Behavior and Social Dominance in Man." *Psychosomatic Medicine* 36:469–47.

Eldredge, Niles, and Marjorie Grene. 1992. *Interactions: The Biological Context of Social Systems*. New York: Columbia University Press.

Eley, Thalia, Paul Lichtenstein, and Jim Stevenson. 1999. "Sex Differences in the Etiology of Aggressive and Non-aggressive Antisocial Behavior: Results from Two Twin Studies." *Child Development* 70 (1): 155–68.

Ellis, Lee, and S. Cole-Harding. 2001. "The effects of prenatal stress, and of prenatal alcohol, on human sexual orientation." *Physiology and Behavior* 74 (1–2): 213–26.

Ellis, Lee, and Jill Hellberg. 2005. "Fetal Exposure to Prescription Drugs and Adult Sexual Orientation." *Personality and Individual Differences* 38 (1): 225–36.

Enç, Berent. 1995. "Units of Behavior." *Philosophy of Science* 62:523–42.

Fagan, Jeffrey. 2004. "Crime, Law, and the Community." In *The Future of Imprisonment*, edited by Michael Tonry. New York: Oxford University Press.

Fagan, Jeffrey, and Garth Davies, 2000. "Street Stops and Broken Windows: *Terry*, Race and Disorder in New York City." *Fordham Urban Law Journal* 28 (2): 457–504.

——— 2004. "The Natural History of Neighborhood Violence." *Journal of Contemporary Criminal Justice* 20:127–47.

Fagan, Jeffrey, V. West, and J. Holland. 2003. "Reciprocal Effects of Crime and Incarceration in New York City Neighborhoods." *Fordham Urban Law Journal* 30:1551–1602.

Fausto-Sterling. 1992. *Myths of Gender*. 2nd ed. New York. Basic Books.

Fay, R. E., C. F. Turner, A. D. Klassen, and J. Gagnon. 1989. "Prevalence and Patterns of Same-Gender Contact among Men." *Science* 243: 338–48.

Feldman, Marcus W. and Richard C. Lewontin. 1975. "The Heritability Hang-Up." *Science* 190:1163–68.

Ferris, Craig, Yvon. Delville, Z. Grzonska, J. Luber-Narod, and T. R. Insel. 1993. "An Iodinated Vasopressin (V_1) Antagonist Blocks Flank Marking and Selectively Labels Neural Binding Sites in Golden Hamsters." *Physiology and Behavior* 54:737–47.

Ferris, Craig, Yvon Delville, Robert Irwin, and Michael Potegal. 1994. "Septo-Hypothalamic Organization of a Stereotyped Behavior Controlled by Vasopressin in Golden Hamsters." *Physiology and Behavior* 55 (4): 755–59.

Ferveur, Jean-Francois, Klemens F. Stortkuhl, Reinhard F. Stocker, Ralph J. Greenspan. 1995. "Genetic Feminization of Brain Structures and Changed Sexual Orientation in Male *Drosophila*." *Science* 267:902–5.

Forastieri, Valter, Cristiane Pinto Andrade, Adriana Lima V. Souza, Monica Santana Silva, Charbel Niño El-hani, Lília Maria De Azevedo Moreira, Luiz Roberto De Barros Mott, and Renato Zamora Flores. 2003. "Evidence Against a Relationship Between Dermatoglyphic Asymmetry and Male Sexual Orientation." *Human Biology* 74 (6): 861–71.

Fowler, Jeanette, Nelly Alia-Klein, Aarti Kiplani, Jean Logan, Benjamin Williams, Wei Zhu, Ian Craig, Frank Telang, Rita Goldstein, Nor Volkow, Paul Vaska, and Gene-Jack Wang. 2007. "Evidence that Brain MAOA Activity Does Not Correspond to MAOA Genotype in Healthy Male Subjects." *Biological Psychiatry* 62:355–58.

Fujimura, Joan. 2006. "Sex Genes: A Critical Sociomaterial Approach to the Politics and Molecular Genetics of Sex Determination." *Signs* 32 (1): 49–82

Garland, David. 2001. *The Culture of Control*. New York: Oxford University Press.

Garnets, Linda D., and Letitia Peplau. 2001. "A New Paradigm for Women's Sexual Orientation: Implications for Therapy." *Women and Therapy* 24 (1–2): 111–21.

Gibbard, Allen. 2001. "Genetic Plans, Genetic Differences, and Violence." in *Genetics and Criminal Violence*, edited by David Wasserman and Robert Wachbroit. New York: Cambridge University Press. pp. 169–98.

Gibbons, Ann. 2004. "Tracking the Evolutionary History of a "Warrior Gene." *Science* 304:818.

Gladue, Brian, Richard Green, and Ronald Hellman. 1984. "Neuroendrocrine Response to Estrogen and Sexual Orientation." *Science* 225:1496–99.

Glicksohn, Joseph. 2002. *The Neurobiology of Criminal Behavior*. New York and Vienna: Springer.

Goff, Philip, Jennifer Eberhardt, Melissa Williams, and Matthew Jackson. 2008. "Not Yet Human: Implicit Knowledge, Historical Dehumanization, and Contemporary Consequences." *Journal of Personality and Social Psychology*. 94 (2): 292–306.

Goldsmith, H. Hill. 1993. "Nature-Nurture Issues in the Behavioral Genetic context: Overcoming Barriers to Communication." in *Nature, Nurture, and Psychology*, edited by Robert Plomin and Gerald McClearn. 325–39. Washington, DC: American Psychological Association.

Goldsmith, H. Hill, Irving I. Gottesman, and K. S. Lemery. 1997. "Epigenetic Approaches to Developmental Psychopathology." *Development and Psychopathology* 9 (2): 365–87.

Gooren, Louis. 1990. "Biomedical Theories of Sexual Orientation: A Critical Examination." In *Homosexuality/Heterosexuality: Concepts of Sexual Orientation*, edited by David McWhirter, Stephanie Sanders, and June Machover Reinisch, 71–81. Oxford: Oxford University Press.

Gottesman, Irving, and H. Hill Goldsmith. 1994. "Developmental Psychopathology of Antisocial Behavior: Inserting Genes into Its Ontogenesis and Epigenesis." In *Threats to Optimal Development: Integrating Biological, Psychological, and Social Risk Factors. Minnesota Symposia on Child Psychology*, vol. 27, edited by Charles Alexander Nelson, 69–104. Hillsdale, NJ: Lawrence Erlbaum Associates.

Gottlieb, Gilbert. 1991. "Experimental Canalization of Behavioral Development: Theory." *Developmental Psychology* 27 (1): 4–13.

———. 1995. "Some Conceptual Deficiences in "Developmental" Behavior Genetics." *Human Development* 38 (3): 131–41.

———. 1997. *Synthesizing Nature-nurture: Prenatal Roots of Instinctive Behavior*. Hillsdale, NJ: Lawrence Erlbaum Associates.

———. 2001. *Individual Development and Evolution: The Genesis of Novel Behavior*. Hillsdale, NJ: Lawrence Erlbaum Associates.

———. 2001a. "A Developmental Psychobiological Systems View: Early Formulation and Current Status." in *Cycles of Contingency*, edited by Susan Oyama, Paul Griffiths, and Russell D. Gray, 41–54. Cambridge, MA: MIT Press.

Gottlieb, Gilbert, Douglas Wahlsten, and R. Lickliter. 2006. "The Significance of Biology for Human Development: A Developmental Psychobiological Systems View." In *Handbook of Child Psychology*. Vol. 1, *Theoretical Models of Human Development*, edited by R. M. Lerner, 233–73. New York: Wiley.

Gottman, John, Neil Jacobson, Regina Rushe, and Joann Shortt. 1995. "The Relationship Between Heart Rate Reactivity, Emotionally Aggressive Behavior, and General Violence in Batterers." *Journal of Family Psychology* 9 (3): 227–48.

Gould, Stephen J. 1981. *The Mismeasure of Man*. New York: W. W. Norton.

Gould, Stephen J., and Richard Lewontin. 1979. "The Spandrels of San Marco and the Panglossian Paradigm." *Proceedings of the Royal Society of London*. B205:581–98.

Granic, Isabela, and Alex V. Lamey. 2002. "Combining Dynamic Systems and Multivariate Analyses to Compare the Mother-Child Interactions of Externalizing Subtypes." *Journal of Abnormal Child Psychology* 30:265–83.

Granic, Isabela, and Gerald R. Patterson. 2006. "Toward a Comprehensive Model of Antisocial Development: A Dynamic Systems Approach." *Psychological Review* 113 (1): 101–31.

Greenberg, Aaron S., and J. Michael Bailey. 2001. "Parental Selection of Children's Sexual Orientation." *Archives of Sexual Behavior* 30 (4): 423–37.

Greene, Joshua, Brian Somerville, Leigh Nystrom, John Darley, and Jonathan Cohen. 2001. "An fMRI Investigation of Emotional Engagement in Moral Judgment." *Science* 293:2105–8.

Greene, Joshua, Leigh Nystrom, Andrew Engell, John Darley, and Jonathan Cohen. 2004. "The Neural Bases of Cognitive Conflict and Control in Moral Judgment." *Neuron* 44 (2): 389–400.

Grene, Marjorie. 1988. "Hierarchies and Behavior." In *Evolution of Social Behavior and Integrative Levels*. T. C. Schneirla Conference Series, vol. 3., edited by Gary Greenberg, Ethel Toback, et. al., 3–17. Hillsdale, NJ: Lawrence Erlbaum Associates.

Griffiths, Paul E. and R. D. Gray. 1994. "Developmental Systems and Evolutionary Explanation." *The Journal of Philosophy* 91 (6): 277–305.

Griffiths, Paul E and Karola Stotz. 2006. "Genes in the Postgenomic Era." *Theoretical Medicine and Bioethics* 27:499 –521.

Grimbos, Teresa, Khytam Dawood, Robert P. Burris, Kenneth J. Zucker; and David A. Puts. 2010. "Sexual Orientation and the Second to Fourth Finger Length Ratio: A Meta-Analysis in Men and Women." *Behavioral Neuroscience* 124 (2): 278–87.

Gross, Matthias. 2004. "Human Geography and Ecological Sociology: The Unfolding of a Human Ecology, 1890–1930—and Beyond." *Social Science History*. 28, 4:575–605.

Guo, Guang, Xiao-Ming Ou, Michael Roettger, and Jean C. Shih. 2008. "The VNTR 2 Repeat in MAOA and Delinquent Behavior in Adolescence and Young Adulthood: Associations and MAOA Promoter Activity." *European Journal of Human Genetics* 16:626–34.

Haapasalo, Jaana and Richard Tremblay. 1994. "Physically Aggressive Boys From Ages 6 to 12: Family Background, Parenting Behavior, and Prediction of Delinquency." *Journal of Consulting and Clinical Psychology* 62 (5): 1044–52.

Hall, J. A. Y. and Doreen Kimura. 1994. "Dermatoglyphic Asymmetry and Sexual Orientation in Men" *Behavioral Neuroscience* 108:1203–6.

Hall, Jeffery. 1994. "The Mating of a Fly." *Science* 264:1702–14.

Hall, Lynn S. and Craig T. Love. 2003. "Finger-Length Ratios in Female Monozygotic Twins Discordant for Sexual Orientation." *Archives of Sexual Behavior* 32 (1): 23–29.

Halperin, Jeffrey, Kathleen McKay, and Jeffrey Newcorn. 2002. "Development, Reliability, and Validity of the Children's Aggression Scale – Parent Version." *Journal of the American Academy of Child and Adolescent Psychiatry* 41 (3): 245–54.

Hamer, Dean. 2006. "Impact of Behavior Genetics on Medicine and Society." Paper presented at the *Interpreting Complexity* Conference sponsored by Center for the Integration of Research on Genetics and Ethics. Stanford University, Stanford, CA. June 6

Hamer, Dean, and Peter Copeland. 1994. *The Science of Desire: The Search for the Gay Gene and the Biology of Behavior*. New York: Simon & Schuster.

Hamer, Dean, S. Hu, V. L. Magnuson. And A. M. Pattatucci. 1993. "A Linkage Between DNA Markers on the X Chromosome and Male Sexual Orientation." *Science* 261:321–27.

Haney, Craig, Curtis Banks, and Philip Zimbardo. 1973. "Interpersonal Dynamics in a Simulated Prison." *International Journal of Criminology and Penology* 1 (1): 69–97.

Harris, Judith. 1999. *The Nurture Assumption.* New York: Free Press.

Harry, Joseph. 1989. "Parental Physical Abuse and Sexual Orientation in Males." *Archives of Sexual Behavior* 18 (3): 251–61.

Haslanger, Sally. 2011. *The Carus Lectures.* Presented at the 2011 meetings of the Pacific Division of the American Philosophical Association, April 20–23, 2011, San Diego, CA. Available from author.

Hawthorne, Susan C. C. 2010. "Institutionalized Intolerance of ADHD: Sources and Consequences." *Hypatia* 25 (3): 504–25.

Haynes, James D. 1995. "A Critique of the Possibility of Genetic Inheritance of Homosexual Orientation." *Journal of Homosexuality* 28 (1–2): 91–113.

Heiligenstein, John, Emil Coccaro, Janet Potvin, Charles Beasley,1992. "Fluoxetine Not Associated With Increased Violence or Aggression in Controlled Clinical Trials." *Annals of Clinical Psychiatry* 4 (4): 285–95.

Henzi, Peter, and Louise Barrett. 2003. "Evolutionary Ecology, Sexual Conflict, and Behavioral Differentiation among Baboon Populations." *Evolutionary Anthropology* 12 (5): 217–30.

Herdt, Gilbert, and Robert Stoller. 1990. *Intimate Communications: Erotics and the Study of Culture.* New York: Cambridge University Press.

Hill, Jonathan. 2002. "Biological, Psychological and Social Processes in the Conduct Disorders." *Journal of Child Psychology and Psychiatry* 43 (1): 133–64.

Hines, Melissa. 2006. "Prenatal Testosterone and Gender-Related Behavior." *European Journal of Endocrinology* 155 (Supplement): S115–21.

Hines, Melissa, Susan Golombok, John Rust, Katie J. Johnston, and Jean Golding. 2002. "Testosterone During Pregnancy and Gender Role Behavior of Preschool Chlidren: A Longitudinal Population Study." *Child Development* 73 (6): 1678–87.

Holden, C. 1992 "Back to the Drawing Board, Says NIH." *Science* 257:1474.

Hubbard, Ruth, and Elijah Wald. 1993. *Exploding the Gene Myth.* Boston: Beacon Press.

Huhman, Kim, Timothy O. Moore, Craig F. Ferris, Edward H. Moogey, and James L. Meyerhoff. 1991. "Acute and Repeated Exposure to Social Conflict in Male Golden Hamsters: Increases in Plasma POMC-Peptides and Cortisol and Decreases in Plasma Testosterone." *Hormones and Behavior* 25:206–16.

Hutchings, Barry, and Sarnoff Mednick. 1975. "Registered Criminality in the Adoptive and Biological Parents of Registered Criminal Adoptees." in *Genetic Research in Psychiatry* edited by R. R. Fieve, D. Rosenthal, H. Brill, 105–16. Baltimore, MD. Johns Hopkins University Press.

Jackson, Jacquelyn Faye. 1993. "Human Behavioral Genetics, Scarr's Theory, and Her Views on Interventions: A Critical Review and Commentary on Their Implications for African American Children." *Child Development* 64:1318–32.

Jacobs, Andrea. 2004. "Prison Power Corrupts Absolutely." *California Western Law Review* 41: 277–301.

Jensen, Arthur. 1969. "How Much Can We Boost I. Q. and Scholastic Achievement?" *Harvard Educational Review* 33:1–123.

Johnson, Jeffrey, Elizabeth Smailes, Patricia Cohen, Stephanie Kasen, and Judith Brook. 2004.

"Anti-Social Parental Behavior, Problematic Parenting and Aggressive Offspring Behaviour During Adulthood." *British Journal of Criminology* 44:915–30.

Johnson, Mark H., ed. 1993. *Brain Development and Cognition.* Oxford, UK: Basil Blackwell.

Jones, Owen D. 2006. "Behavioral Genetics and Crime, in Context." *Law and Contemporary Problems* 68:81–100.

Kaplan, Jonathan. 2000. *The Limits and Lies of Human Genetic Research.* New York: Routledge.

Kavoussi, Richard, and Emil Coccaro. 1993. "The Amphetamine Challenge Test Correlates With Affective Lability in Healthy Volunteers." *Psychiatry Research* 48:219–28.

Kavoussi, Richard, Jennifer Liu, and Emil Coccaro. 1994. "An Open Trial of Sertraline in Personality Disordered Patients With Impulsive Aggression." *Journal of Clinical Psychiatry* 55 (4): 137–41.

Keller, Evelyn F. 2005. "DDS: Dynamics of Developmental Systems." *Biology and Philosophy* 20 (2–3): 409–16.

Kellert, Stephen, Helen Longino, and C. K. Waters. 2006. "The Pluralist Stance." in *Scientific Pluralism*, edited by Stephen Kellert, Helen Longino, and C. K. Waters. vii–xxix. Minneapolis, MN: University of Minnesota Press.

Kelling, George, and Catherine Cole. 1996. *Fixing Broken Windows: Restoring Order and Reducing Crime in Our Communities.* New York. Simon and Schuster.

Kendal, Jeremy, Marcus Feldman, and Kenichi Aoki. 2006. "Cultural Co-evolution of Norm Adoption and Enforcement When Punishers Are Rewarded or Non-punishers Are Punished." *Journal of Theoretical Population Biology.* 70 (1): 10–25.

Kendler, Kenneth. 2008. "Explanatory Models for Psychiatric Illness." *American Journal off Psychiatry* 165" 695–702.

Kendler, Kenneth, and J. Campbell. 2009 "Interventionist Causal Models in Psychiatry: Repositioning the Mind-Body Problem." *Psychological Medicine* 39:882–87.

Kendler, Kenneth, Charles Gardner, and Carol Prescott. 2002. "Toward a Comprehensive Developmental Model for Major Depression in Women." *American Journal of Psychiatry* 159:1133–1145.

Kevles, Daniel, and LeRoy Hood, eds. 1993. *The Code of Codes.* Cambridge, MA: Harvard University Press.

Kiehl, Kent. 2006. "A Cognitive Neuroscience Perspective on Psychopathy: Evidence for Paralimbic System Dysfunction." *Psychiatry Research* 142, 2:107–28.

Kiehl, Kent, Andra M. Smith, Robert D. Hare, Adrianna Mendreck, Bruce B. Foster, Johann Brink, Peter F. Liddle. 2001. "Limbic Abnormalities in Affective Processing by Criminal Psychopaths as Revealed by Functional Magnetic Resonance Imaging." *Biological Psychiatry* 50 (9): 677–84.

King, Mary-Claire. 1993. "Sexual Orientation and the X." *Nature* 364 (6435): 288–89.

Kindlon, Daniel, Richard Tremblay, Enrico Mezzacappa, Felton Earls, Denis Laurent, and Benoist Schaal. 1995. "Longitudinal Patterns of Heart Rate and Fighting Behavior in 9- through 12-Year-Old Boys." *Journal of the American Academy of Child and Adolescent Psychiatry* 34:371–77.

Kitamoto, Toshihiro. 2002. "Conditional Disruption of Synaptic Transmission Induces Male-Male Courtship in *Drosophila.*" *Proceedings of the National Academy of Sciences* 99 (20): 13232–37.

Kitcher, Philip. 2002. *Science, Truth, and Democracy.* New York: Oxford University Press.

Klonsky, E. David, Thomas Oltmanns, and Eric Turkheimer. 2002. "Informant Reports of Personality Disorder: Relation to Self-Report, and Future Research Directions." *Clinical Psychology: Science and Practice* 9:300–311.

Knorr-Cetina, Karin. 1981. *The Manufacture of Knowledge*. Oxford: Pergamon Press.

Koshland, Daniel. 1989. "Sequences and Consequences of the Human Genome Project" (editorial). *Science* 246:189.

Kreek, Mary Jeanne, David A Nielson, Eduardo R. Butelman, and K. Steven LaForge. 2005. "Genetic Influences on Impulsitivy, Risk Taking, Stress Responsivity, and Vulnerability to Drug Abuse and Addiction." *Nature Neuroscience* 8:1450–57.

Krimsky, Sheldon. 2003. *Science in the Private Interest: Has the Lure of Profits Corrupted Biomedical Research?* Lanham, MD: Rowman and Littlefield.

Landolt, Monica A., Kim Bartholomew, Colleen Saffrey, Doug Oram, and Daniel Perlman. 2004. "Gender Nonconformity, Childhood Rejection, and Adult Attachment: A Study of Gay Men." *Archives of Sexual Behavior* 33 (2): 117–28

Lavigueur, Suzanne, Richard Tremblay, and Jean-Francois Saucier. 1995. "Interactional Processes in Families with Disruptive Boys: Patterns of Direct and Indirect Influence." *Journal of Abnormal Child Psychology* 23 (3): 359–78.

Lee, Cheryl, and Lawrence Morse. 2004. "African-American Employment and Job Quality: Income, Wealth and Health Insurance Benefits." *NUL Quarterly Jobs Report (QJR-04–2004)*. Washington, DC: National Urban League Institute for Opportunity and Equality.

Lee, Royce, and Emil Coccaro. 2001. "The Neuropsychopharmacology of Criminality and Aggression." *Canadian Journal of Psychiatry* 46:35–44.

Lerner, Richard. 1991. "Changing Organism-Context Relations as the Basic Process of Development: A Developmental Contextual Perspective." *Developmental Psychology* 27:27–32.

———. 1998. "Adolescent Development: Challenges and Opportunities for Research, Programs, and Policies." *Annual Review of Psychology.* 49:413–46.

LeVay, Simon. 1991. "A Difference in Hypothalamic Structure Between Heterosexual and Homosexual Men." *Science* 253:1034–37.

LeVay, Simon, and Dean H. Hamer. 1994. "Evidence for a Biological Influence in Male Homosexuality." *Scientific American* 270:44–49.

Levitis, Daniel A., William Z. Lidicker, Jr,, and Glenn Freund. 2009. "Behavioural Biologists Do Not Agree on What Constitutes Behaviour." *Animal Behaviour.* 78:103–110

Lewontin, Richard. 1974. "The Analysis of Variance and the Analysis of Causes." *American Journal of Genetics,* 26:400–411.

———. 1991. *Biology as Ideology: The Doctrine of DNA.* Concord, Ontario: Anansi.

Lewontin, Richard, Steven Rose, and Leon Kamin. 1984. *Not in Our Genes: Biology, Ideology, and Human Nature.* Harmondsworth: Penguin Books.

Liu, Yeuyi I., Paul H. Wise, and Atul Butte. 2009. "The 'Etiome': Identification and Clustering of Human Disease Etiological Factors." *BMC Bioinformatics* 10, Suppl 2: S14.

Lloyd, Elisabeth Anne. 1994. *The Structure and Confirmation of Evolutionary Theory.* Princeton, NJ. Princeton University Press

———. 2001. "Science Gone Astray: Evolution and Rape." *Michigan Law Review* 99 (6): 1536–59.

———. 2005. *The Case of the Female Orgasm: Bias in the Science of Evolution.* Cambridge, MA. Harvard University Press.

———. 2010. "Confirmation and Robustness of Climate Models." *Philosophy of Science* 77: 971–84.

Longino, Helen E. 1990. *Science as Social Knowledge*, Princeton, NJ: Princeton University Press.

———. 2001a. "What Do We Measure When We Measure Aggression?" *Studies in History and Philosophy of Science* 32 (4): 685–701.

———. 2001b. *The Fate of Knowledge*, Princeton, NJ: Princeton University Press.

———. 2002. "Behavior as Affliction: Common Frameworks of Behavior Genetics and Its Rivals." in *Medical Genetics: Conceptual Foundations and Classic Questions*, edited by Rachel Ankeny and Rachel Parker. Boston: Kluwer.

———. 2006. "Theoretical Pluralism and the Scientific Study of Behavior." In *Scientific Pluralism*, edited by Stephen Kellert, Helen Longino, and C. K. Waters. Minneapolis: University of Minnesota Press.

Luntz, Barbara and Cathy Widom. 1994. "Antisocial Personality Disorder in Abused and Neglected Children Grown Up." *American Journal of Psychiatry* 151 (5): 670–74.

Mabry, Patricia, D. H. Olster, G. D. Morgan, and D. B. Abrams. 2008. "Interdisciplinarity and Systems Science to Improve Population Health: A View from the NIH Office of Behavioral and Social Sciences Research." *American Journal of Preventive Medicine* 35 (2): 211–24.

Maccoby, Eleanor. 2000. "Parenting and its Effects on Children: On Reading and Misreading Behavior Genetics." *Annual Reviews in Psychology*. 51 (1): 1–27.

Manuck, S. B., J. D. Flory, R. E. Ferrell, K. M. Dent, J. J. Mann, and M. F. Muldoon. 1999. "Aggression and Anger-related Traits Associated with a Polymorphism of the Tryptophan Hydroxylase Gene." *Biological Psychiatry* 45:603–14.

Manuck S. B., J. D. Flory, R. E. Ferrell, J. J. Mann, and M. F. Muldoon . 2000. "A Regulatory Polymorphism of the Monoamine Oxidase-A Gene may be Associated with Variability in Aggression, Impulsivity, and Central Nervous System Serotonergic Responsivity." *Psychiatry Research* 95 (1): 9–23.

Mason, Dehryl and Paul Frick. 1994. "The Heritability of Antisocial Behavior: A Meta-Analysis of Twin and Adoption Studies." *Journal of Psychopathology and Behavioral Assessment* 16 (4): 301–23.

McClellan, Jay, and Mary-Claire King. 2010. "Genomic Analysis of Mental Illness: A Changing Landscape." *Journal of the American Medical Association*. 303 (24): 2523–24.

McClure, Samuel, David Laibson, George Loewenstein, and Jonathan Cohen. 2004. "Separate Neural Systems Value Immediate and Delayed Monetary Rewards." *Science* 306:503–7.

McCord, Joan. 2001. "Forging Criminals in the Family." In *Handbook of Youth and Justice*, edited by S. O. White, 223–35. New York: Plenum.

McCord, Joan, Richard Tremblay, Frank Vitaro, and Lyse Desmarais-Gervais. 1994. "Boys' Disruptive Behaviour, School Adjustment, and Delinquency: The Montreal Prevention Experiment." *International Journal of Behavioral Development* 17 (4): 739–52.

McGinn, Colin. 2011. "Can the Brain Explain Your Mind?" review of V. S. Ramachandran, *The Tell-Tale Brain: A Neuroscientist's Quest for What makes Us Human* (Norton, 2011). *New York Review of Books* 58 (March 24): 32–35.

McGue, Matt. 1994. "Why Developmental Psychology Should Find Room for Behavioral Genetics." In *Threats To Optimal Development: Integrating Biological, Psychological, and Social Risk Factors*, edited by Charles Nelson Alexander, 105–19. Hove, England: Lawrence Erlbaum Associates, Inc.

McGue, Matt, Steven Bacon, and David Lykken. 1993. "Personality Stability and Change in Early Adulthood: A Behavioral Genetic Analysis." *Developmental Psychology* 29 (1): 96–109.

McGue, Matt, and Thomas J. Bouchard Jr. 1998. "Genetic and Environmental Influences on Human Behavioral Differences." *Annual Review of Neuroscience.* 21 (1998): 1–24.

McWhirter, David, Stephanie Sanders, and June Machover Reinisch. 1990a. "Homosexuality/ Heterosexuality: An Overview." In *Homosexuality/Heterosexuality: Concepts of Sexual Orientation*, edited by David McWhirter, Stephanie Sanders, and June Machover Reinisch, xix–xxvii. Oxford: Oxford University Press.

———, eds. 1990b. *Homosexuality/Heterosexuality: Concepts of Sexual Orientation.* Oxford: Oxford University Press.

Mealey, Linda. 1995. "The Sociobiology of Sociopathy: An Integrated Evolutionary Model." *Behavioral and Brain Sciences* 18 (3): 523–99.

Mejia, Jose Maria, Frank Ervin, Glen B. Baker, Roberta M. Palmour. 2002. "Monoamine Oxidase Inhibition during Brain Development Induces Pathological Aggressive Behavior in Mice." *Biological Psychiatry* 52:811–21.

Meyer-Bahlburg, Hermann. 1977. "Sex Hormones and Male Homosexuality in Comparative Perspective." *Archives of Human Behavior* 6 (4): 297–325.

Miles, Donna, and Gregory Carey. 1997. "Genetic and environmental architecture of human aggression." *Journal of Personality and Social Psychology.* 72 (1): 207–17

Milgram, Stanley. 1974. *Obedience to Authority: An Experimental View.* London: Tavistock.

Miller, Greg. 2010. "The Seductive Allure of Behavioral Epigenetics." *Science* 329:24–27

Miller-Johnson, Shari, John D. Coie, Anne Maumary-Geraud, and Karen Bierman. 2002. "Peer Rejection and Aggression and Early Starter Models of Conduct Disorder." *Journal of Abnormal Child Psychology* 30 (3): 217–31.

Mills, Shari, and Adrian Raine. 1994. "Neuroimaging and Aggression." *Journal of Offender Rehabilitation* 23 (3–4): 145–58.

Mitchell, Sandra. 2002. "Integrative Pluralism." *Biology and Philosophy* 17:55–70.

———. 2009. "Explaining Complex Behavior." In *Philosophical Issues in Psychiatry: Explanation, Phenomenology, and Nosology*, edited by Kenneth S. Kendler and Josef Parnas. Baltimore: Johns Hopkins Unversity Press.

Moffitt, Terrie E. 2005a. "Genetic and Environmental Influences on Antisocial Behaviors: Evidence from Behavioral-Genetic Research." *Advances in Genetics* 55:41–104.

———. 2005b. "The New Look of Behavioral Genetics in Developmental Psychopathology: Gene-Environment Interplay in Antisocial Behaviors." *Psychological Bulletin.* 131 (4): 533–54.

Moffitt, Terrie E., Avshalom Caspi, Julia Kim-Cohen, and Alan Taylor. 2004. "Genetic and Environmental Processes in Young Children's Resilience and Vulnerability to Socioeconomic Deprivation." *Child Development.* 75 (3): 651–68.

Moffitt, Terrie E., Avshalom Caspi, and Michael Rutter. 2005. "Strategy for Investigating Interactions Between Measured Genes and Measured Environments." *Archives of General Psychiatry* 62 (2): 473–81.

Money, John. 1990. "Agenda and Credenda of the Kinsey Scale." In *Homosexuality/Heterosexuality: Concepts of Sexual Orientation*, edited by David McWhiter, Stephanie Sanders, and June Machover Reinisch, 41–60. Oxford: Oxford University Press.

Muhammad, Fida, Michael Shaughnessy, Scott Johnson, and Russell Eisenman. 2002. "Understanding Torture and Torturers." *Journal of Evolutionary Psychology* 23 (3–4): 131–48.

Mustanski, Brian, J. Michael Bailey, and Sarah Kaspar. 2002. "Dermatoglyphics, Handedness, and Sexual Orientation." *Archives of Sexual Behavior* 31 (1): 113–22.

Mustanski, Brian, Michael DePree, Caroline Nievergelt, Sven Bocklandt, Nicholas Schork, Dean Hamer. 2005. "A Genomewide Scan of Male Sexual Orientation." *Nature Genetics* 116:272–78.

National Science Foundation. 2009. *Women, Minorities, and Persons with Disabilities in Science and Engineering 2009*. NSF 09–305. Arlington, VA. Available from http://www.nsf.gov/statistics/wmpd/.

Nelkin, Dorothy, and Susan Lindee. 1995. *The DNA Mystique*. New York: Freeman.

New, Antonia S. Joel Gelernter, Yoram Yovell, Robert Trestman, David Nielsen, Jeremy Silverman, Vivian Metropoulou, and Larry Siever. 1998. "Tryptophan Hydroxylase Genotype is Associated with Impulsive-Aggression Measures." *American Journal of Medical Genetics (Neuropsychiatric Genetics)* 81:13–17.

Olivier, Berend. 2004. "Serotonin and Human Aggression." *Annals of the New York Academy of Sciences*. 1036:382–92.

Olivier, Berend, and Ruud van Oorschot. 2005. "5HT$_{1B}$ Receptors and Aggression: A Review." *European Journal of Pharmacology* 526 (1–3): 207–17.

Orpinas, Francis, and Ralph Frankowski. 2001. "The Aggression Scale: A Self-Report Measure of Aggressive Behavior for Young Adolescents." *Journal of Early Adolescence* 21 (1): 50–67.

Ortner, Sherry, and Harriet Whitehead. 1981. "Introduction: Accounting for Sexual Meanings." In *Sexual Meanings*, edited by Sherry Ortner and Harriet Whitehead, 1–27. Cambridge: Cambridge University Press.

Ostrom, Elinor. 2007. "Challenges and Growth: The Development of the Interdisciplinary Field of Institutional analysis." *Journal of Institutional Economics* 3 (3): 239–64.

Oyama, Susan. 1985. *The Ontogeny of Information*. New York, NY: Cambridge University Press.

———. 2000. *Evolution's Eye: A Systems View of the Biology-Culture Divide*. Durham, NC: Duke University Press.

Palca, J. 1992. "NIH Wrestles with Furor over Conference." *Science* 257:739.

Palomaki, Glenn E., and James E. Haddow. 1993. "Is It Time for Population-Based Prenatal Screening for Fragile-X?" *Lancet* 341:373–74.

Peplau, Letitia Anne. 2001. "Rethinking Women's Sexual Orientation: An Interdisciplinary, Relationship-focused Approach." *Personal Relationships* 8:1–19.

Peplau, Letitia Anne, and Linda Garnets. 2000. "A New Paradigm for Understanding Women's Sexuality and Sexual Orientation." *Journal of Social Issues* 56 (2): 329–50.

Pettit, Becky. 2009. "Enumerating Inequality: The Constitution, the Census Bureau, and the Criminal Justice System." *Connecticut Public Interest Law Journal* 5 (2): 139–66.

Pettit, Becky, and Bruce Western. 2004. "Mass Imprisonment and the Life Course: Race and Class Inequality in U.S. Incarceration." *American Sociological Review* 69 (2): 151–69.

Pfaff, Donald. 1980. *Estrogen and Brain Function*. New York: Springer Verlag.

Pillard, Richard C., and James Weinrich. 1986. "Evidence of Familial Nature of Male Homosexuality." *Archives of General Psychiatry* 46:808–12.

Pinker, Stephen. 2002. *The Blank Slate*. New York, NY: Viking Penguin.

Plaisance, Kathryn. 2006. "Behavioral Genetics and the Environment: The Generation and Exportation of Scientific Claims." PhD diss., University of Minnesota.

Plöderl, Martin, and Reinhold Fartacek. 2009 "Childhood Gender Nonconformity and Harassment as Predictors of Suicidality among Gay, Lesbian, Bisexual and Heterosexual Austrians." *Archives of Sexual Behavior* 38 (3): 400–410.

Plomin, Robert, Terryl T. Foch, and David C. Rowe. 1981. "Bobo Clown Aggression in Childhood: Environment, Not Genes." *Journal of Research in Personality* 15 (2): 331–42.

Plomin, Robert, Michael Owen, and Peter McGuffin. 1994. "The Genetic Basis of Complex Human Behavior." *Science* 264:1733–39.

Rahman, Qazi. 2005. "The Neurodevelopment of Human Sexual Orientation." *Neuroscience and Biobehavioral Reviews* 29 (7): 1057–66.

Rahman, Qazi, and G. D. Wilson. 2003. "Sexual Orientation and the 2nd to 4th Finger Length Ratio: Evidence for Organising Effects of Sex Hormones or Developmental Instability?" *Psychoneuroendocrinology* 28 (3): 288–303.

Raine, Adrian. 2008. "From Genes to Brain to Antisocial Behavior." *Current Directions in Psychological Science*. 17 (5): 323–28.

Raine, Adrian, Monte Buchsbaum, Jill Stanley, Steven Lottenberg, Leonard Abel and Jacqueline Stoddard. 1994. "Selective Reductions in Prefrontal Glucose Metabolism in Murders." *Biological Psychiatry* 36 (6): 365–73.

Raine, Adrian, Patricia Brennan, and David Farrington. 1997. "Biosocial Bases of Violence: Conceptual and Theoretical Issues." In *Biosocial Bases of Violence*, edited by Adrian Raine, Patricia Brennan, David Farrington, and Sarnoff Mednick, 1–21. New York: Plenum Press.

Raine, Adrian, Patricia Brennan, and Sarnoff Mednick. 1994. "Birth Complications Combined with Early Maternal Rejection at Age 1 Year Predispose to Violent Crime at Age 18 Years." *Archives of General Psychiatry* 51:984–88.

Ramachandran, V. S., and Colin McGinn. 2011. "The Tell-Tale Brain: An Exchange." *New York Review of Books* 58 (June 23): 68.

Rasmussen, Nicholas. 1997. *Picture Control: The Electron Microscope and the Transformation of Biology in America, 1940–1960.* Palo Alto, CA: Stanford University Press.

Rhee, Soo Hyun, and Irwin D. Waldman. 2002. "Genetic and Environmental Influences on Antisocial Behavior: A Meta-Analysis of Twin and Adoption Studies." *Psychological Bulletin* 128 (3): 490–529.

Rice, George, Carol Anderson, Neil Risch, and George Ebers. 1999. "Male Homosexuality: Absence of Linkage to Microsatellite Markers at Xq28." *Science* 284:665–67.

Richardson, Robert. 1984. "Biology and Ideology: The Interpenetration of Science and Values" *Philosophy of Science* 51 (3): 396–420.

Richardson, Sarah. 2009. *Gendering the Genome.* PhD Diss. Stanford University.

Risch, Neil, Richard Herrell, Thomas Lehner, Kung-Yee Liang, Lindon Eaves, Josephine Hoh, Andrea Griem, Maria Kovacs, Jurg Ott, and Kathleen Ries Merikangas. 2009. "Interaction between the Serotonin Transporter Gene 5-HTTLPR, Stressful Life Events, and Risk of Depression: A Meta-analysis." *Journal of the American Medical Association* 301 (23): 2462–71.

Rivera, Beverly and Cathy Widom. 1990. "Childhood Victimization and Violent Offending." *Violence and Victims* 5 (1): 19–35.

Robinson, Matthew. 2009. "No Longer Taboo: Crime Prevention Implications of Biosocial Criminology." In *Biosocial Criminology*, edited by Anthony Walsh and Kevin Beaver, 243–63. New York: Routledge.

Rose, Nikolas. 2000. "The Biology of Culpability: Pathological Identity and Crime Control in a Biological Culture." *Theoretical Criminology*, 4:5–34

Rose, Richard. 1995. "Genes and Human Behavior." *Annual Review of Psychology* 46:625–54.

Roush, Wade. 1995. "Conflict Marks Crime Conference." *Science* 269:1808–9.

Rowe, David C., and Robert Plomin. 1984. "The Importance of Nonshared (E_1) Environmental Influences in Behavioral Developmental Development." *Developmental Psychology* 17:17–31.

Sabol, Sue Z., Stella Hu, Dean Hamer. 1998. "A Functional Polymorphism in the Monoamine Oxidase Gene Promoter." *Human Genetics* 103:273–79.

Sampson, Robert J., Stephen W. Raudenbush, and Felton Earls. 1997. "Neighborhoods and Violent Crime: A Multilevel Study of Collective Efficacy." *Science* 277:918–24.

Sanfey, Alan, James Rilling, Jessica Aronson, Leigh Nystrom, and Jonathan Cohen. 2003. "The Neural Basis of Decision-Making in the Ultimatum Game." *Science* 300:1755–58.

Sapolsky, Robert. 2005. "The Influence of Social Hierarchy on Primate Health." *Science* 308:648–52.

Scarr, Sandra. 1987. "Three Cheers for Behavior Genetics: Winning the War and Losing Our Identity." *Behavior Genetics* 17 (3): 219–28.

———. 1992. "Developmental Theories for the 1990s: Development and Individual Differences." *Child Development* 63 (1): 1–19.

———. 1993. "Biological and Cultural Diversity: The Legacy of Darwin for Development." *Child Development* 64:1333–53.

———. 1994. "Why Developmental Research Needs Evolutionary Theory: To Ask Interesting Questions." In *International Perspectives on Psychological Science*, vol. 1, *Leading Themes*, edited by Paul Bertelson and Paul Helen, 170–78. Hove, England: Lawrence Erlbaum Associates.

———. 1995. "Commentary." *Human Development* 38:154–57.

———. 1997. "Behavior-Genetic and Socialization Theories of Intelligence." In *Intelligence, Heredity, and Environment*, edited by Robert Sternberg and Elena Grigorenko, 3–41. New York, NY: Cambridge University Press.

Scarr, Sandra, and Kathleen McCartney. 1983. "How People Make Their Own Environments: A Theory of Genotype $\rightarrow$ Environment Effects." *Journal of Personality and Social Psychology* 40 (5): 674–92

Schaffner, Kenneth. 1998. "Genes, Behavior, and Developmental Emergentism: One Process, Indivisible?" *Philosophy of Science* 65 (2): 276–89.

Schüklenk, Udo. 1993. "Is Research into the Cause(s) of Homosexuality Bad for Gay People?" *Christopher Street*. 208:13–15.

Searle, John. 1992. *The Rediscovery of the Mind*. Cambridge MA. MIT Press.

———. 2011. "The Mystery of Consciousness Continues." Review of Antonio Damasio, *Self Comes to Mind: Constructing the Conscious Brain* (Pantheon, 2011). *New York Review of Books* 58 (June 9): 50–52.

Sen, Amartya. 1990. "More Than 100 Million Women Are Missing." *New York Review of Books* 37 (December 20). http://www.nybooks.com/articles/3408 (accessed July 7, 2009).

Sesardic, Neven. 1993. "Heritability and Causality." *Philosophy of Science* 60 (3): 396–418.

———. 2000. " Philosophy of Science that Ignores Science: Race, I.Q. and Heritability." *Philosophy of Science* 67 (4): 580–602.

————. 2003. "Heritability and Indirect Causation." *Philosophy of Science* 70 (5): 1002–14.

Shuldiner, Alan R. and Toni Pollin. 2010. "Variations in Blood Lipids." *Nature* 466 (5): 703–4

Simm, Michael. 1994. "Violence Study Hits a Nerve in Germany" *Science* 264:253.

Simon, Neal G. and Shi-Fang Lu. 2006. "Androgens and Aggression" in *Biology of Aggression,* edited by Randy J. Nelson, 211–30. New York: Oxford University Press.

Sluyter F., J. N. Keisjer, D. I. Boomsma, L. J. P. van Doornen, E. J. C. G. van den Oord, and H. Sneider. 2000. "Genetics of Testosterone and the Aggression-Hostility-Anger (AHA) Syndrome: a Study of Middle-aged Male Twins." *Twin Research* 3:266–276.

Smith, Linda, Esther Thelen, Robert Titzer, and Dewey McLin. 1999. "Knowing in the Context of Acting: The Task Dynamics of the A-Not-B Error" *Psychological Review* 106 (2): 235–60.

Sober, Elliott. 2001. "Separating Nature and Nurture." In *Genetics and Criminal Behavior* edited by David Wasserman and Robert Wachbroit, 47–78. Cambridge, UK and New York: Cambridge University Press.

Speltz, Matthew, Michelle De Kleyn, Mark T. Greenberg, and Marilyn Dryden.1995. "Clinic Referral for Oppositional Defiant Disorder: Relative Significance of Attachment and Behavioral Variables." *Journal of Abnormal Child Psychology* 23 (4): 487–507.

Spencer, Quayshawn. 2009. "A Critique of the Cladistic Race Concept." PhD diss., Stanford University.

Staner, Luc, Gökhan Uyanik, Humberto Correa, Fabien Tremeau, Jose Monreal, Marc-Antoine Crocq, Grigori Stefos, Deborah J. Morris-Rosendahl, Jean-Paul Macher. 2002. "A Dimensional Impulsive-Aggressive Phenotype is Associated with the A218C Polymorphism of the Tryptophan Hydroxylase Gene." *American Journal of Medical Genetics* 114 (5): 553–57.

Steele, Claude M. 1997. "A Threat in the Air: How Stereotypes Shape the Intellectual Identities and Performance of Women and African Americans." *American Psychologist* 52, 613–29.

Steele, Claude M., and Aronson, Joshua A. 2004 "Stereotype Threat Does Not Live by Steele and Aronson alone." *American Psychologist* 59:47–48.

Stein, Edward, ed. 1992. *Forms of Desire: Sexual Orientation and the Social Construction Controversy.* New York: Routledge.

Stern, Vivian. 1997. *A Sin Against the Future.* Boston, MA: Northeastern University Press.

Stone, Richard. 1992 "HHS 'Violence Initiative' Caught in a Crossfire." *Science* 258:212–3.

Stotz, Karola. 2006. "Molecular Epigenesis: Distributed Specificity as a Break in the Central Dogma." *History and Philosophy of the Life Sciences* 28 527–44.

————. 2008. "The Ingredients for a Postgenomic Synthesis of Nature and Nurture." *Philosophical Psychology* 21,3:359 -381.

Sukhodolsky, D. G., and V. Ruchkin. 2006. "Evidence-based Psychosocial Treatments in the Juvenile Justice System." *Child and Adolescent Psychiatric Clinics of North America.* 15 (2): 501–16.

Suppes, Patrick, Bing Han, Julie Epelboim, and Zhong-Lin Lu. 1996 "Invariance Between Subjects of Brain Wave Representations of Language." *Proceedings of the National Academy of Sciences* 96 (22): 12953–58.

————. 1999 "Invariance of Brain-Wave Representation of Simple Visual Images and Their Names." *Proceedings of the National Academy of Sciences* 96 (25): 14658–63.

Tabery, James. 2009. "Making Sense of the Nature-Nurture Debate." *Biology and Philosophy* 24:711–23.

Taylor, Kenneth. 2001. "On the Explanatory Limits of Behavioral Genetics." In *Genetics and Crimi-*

nal Behavior, edited by David Wasserman and Robert Wachbroit, 117–39. Cambridge: Cambridge University Press.

Tellegen, Auke, David T. Lykken, Thomas J. Bouchard, Jr., Kimerly J. Wilcox, Nancy .L. Segal, and Stephen Rich. 1988. "Personality Similarity in Twins Reared Apart." *Journal of Personality and Social Psychology* 54 (6): 1031–39.

Thelen, Esther. 2000. "Grounded in the World: Developmental Origins of the Embodied Mind." *Infancy* 1 (1): 3–28.

Touchette, Nancy. 1995. "Genetics and Crime Conference Reaps Subtle Benefit." *Nature Medicine* 1 (11): 1108–9.

Trout, J. D. 1998. *Measuring the Intentional World*. New York, NY: Oxford University Press.

Turkheimer, Eric. 2008. "More Alike than Different: Genetic and Environmental Explanations of Behavior" Lecture at Biological Explanations of Behavior Conference, Hannover Germany, June 12–15.

———. 2009. "Causation in Human Behavioral Genetics." Symposium Presentation at Society for the Philosophy of Science in Practice Conference, Minneapolis, MN. June 17–20.

———. 2012. "Genome Wide Association Studies of Behavior Are Social Science." In *Philosophy of Behavioral Biology: Boston Studies in the Philosophy of Science*, vol. 282, edited by Kathryn Plaisance and Thomas Reydon, 43–64. Springer.

Turkheimer, Eric, H. Hill Goldsmith, and Irving Gottesman. 1995. "Some Conceptual Deficiencies in 'Developmental' Behavior Genetics: Comment." *Human Development* 38 (3): 143–53.

Turkheimer, Eric, and Irving Gottesman. 1991. "Individual Differences and the Canalization of Human Behavior." *Developmental Psychology* 27 (1): 18–22.

US Department of Justice. 2011. *The NIJ Conference 2011*. Washington, DC: National Institute of Justice. http://www.nij.gov.

van Fraassen, Bas. 2008. *Scientific Representation*. Oxford: Oxford University Press.

———. 2009. "Perils of Perrin." *Philosophical Studies* 143:25–32

Van Wyck, Paul H., and Chrisann S. Geist. 1984. "Psychosocial Development of Heterosexual, Homosexual, and Bisexual Behavior." *Archives of Sexual Behavior* 13 (6): 505–44.

Villaroel, Maria, Charles F. Turner, Elizabeth Eggleston, Alia Al-Tayyib, Susan M. Rogers, Anthony M. Roman, Philip C. Cooley and Harper Gordek. 2006. "Same-Gender Sex in the United States: Impact of T-Acasi on Prevalence Estimates." *Public Opinion Quarterly* 70 (2): 166–96. doi: 10.1093/poq/nfj023

Wade, Nicholas. 2007. "In the Genome Race, the Sequel is Personal." *The New York Times* (September 4, 2007): D1.

Wahlsten, Douglas, and Gilbert Gottlieb. 1997. "The Invalid Separation of Effects of Nature and Nurture: Lessons From Animal Experimentation." In *Intelligence, Heredity, and Environment*, edited by Robert Sternberg and Elena Grigorenko, 163–92. Cambridge: Cambridge University Press.

Walsh, Anthony, and Kevin Beaver. 2009. *Biosocial Criminology*. New York: Routledge.

Wasserman, David, and Robert Wachbroit. 2001. "Introduction." In *Genetics and Criminal Behavior*, edited by David Wasserman and Robert Wachbroit. Cambridge: Cambridge University Press.

Waters, C. Kenneth. 2004."What Was Classical Genetics?" *Studies in History and Philosophy of Science A* 35:783–809

———. 2006. "A Pluralist Interpretation of Gene-Centered Biology," In *Scientific Pluralism: Min-*

nesota Studies in the Philosophy of Science Volume XIX, edited by Stephen Kellert, Helen Longino, and Waters, eds.. Minneapolis: Minnesota UP. pp. 190–214,

———. 2007. "Causes That Make a Difference." *Journal of Philosophy* 104 (11): 551–79.

Weinrich, James D. 1995. "Biological Research on Sexual Orientation: A Critique of the Critics." *Journal of Homosexuality* 28 (1–2): 197–213.

Weinshenker, Naomi, and Allan Siegel. 2002. "Bimodal Classificatioan of Aggression: Affective Defense and Predatory Attack." *Aggression and Violent Behavior* 7:237–50.

Weisstein, Naomi. 1971. "Psychology Constructs the Female; or The Fantasy Life of the Male Psychologist (With Some Attention to the Fantasies of His Friends, the Male Biologist, and the Male Anthropologist)." *Social Education* 35 (4): 362–73.

West, D. J. 2001. "Parental Selection of Children's Sexual Orientation: A Commentary." *Archives of Sexual Behavior* 30 (4): 439–41.

West, Meredith, David King and David White. 2003. "The Case for Developmental Ecology." *Animal Behavior* 66:617–22.

Whalen, Richard, David Geary, and Frank Johnson. 1990. "Models of Sexuality." In *Homosexuality/ Heterosexuality: Concepts of Sexual Orientation*, edited by McWhirter, Sanders, and Reinisch, 61–70. Oxford: Oxford University Press.

Whitam, Frederick L., Milton Diamond, and James Martin. 1993. "Homosexual Orientation in Twins: A Report on 61 Pairs and Three Triplet Sets." *Archives of Sexual Behavior* 22 (3): 187–205.

Widom, Cathy. 1989a. "Child Abuse, Neglect, and Adult Behavior: Research Design and Findings on Criminality, Violence, and Child Abuse." *American Journal of Orthopsychiatry* 59:355–67.

———. 1989b. "Does Violence Beget Violence?" *Psychological Bulletin* 106 (1): 3–28.

———. 1989c. "The Intergenerational Transmission of Violence." In *Pathways of Criminal Violence*, edited by Neil Alan Weiner and Marvin E Wolfgang, 137–201. Newbury Park, CA: SAGE Press.

———. 1991. "A Tail on an Untold Tale: Response to 'Biological and Genetic Contributors to Violence: Widom's Untold Tale.'" *Psychological Bulletin* 109 (1): 130–32.

Williams, Terrance, Michelle Pepitone, Scott Christensen, Bradley Cooke, Andrew Huberman, Nicholas Breedlove, Tessa Breedlove, Cynthia Jordan, Marc Breedlove. 2000. "Finger Length Patterns Indicate an Influence of Fetal Androgens on Human Sexual Orientation." *Nature* 404:455–56.

Wingfield, John C. 2005. "A Continuing Saga: The Role of Testosterone in Aggression." *Hormones and Behavior* 48 (3): 253–55.

Yeh, Shih-Rung, Russell A. Fricke, and Donald Edwards. 1996. "The Effect of Social Experience on Serotonergic Modulation of the Escape Circuit of Crayfish." *Science* 271:366–69.

Zhang, Shang-Ding, and Ward Odenwald. 1995. "Misexpression of the White (*w*) Gene Triggers Male-Male Courtship in *Drosophila*." *Proceedings of the National Academy of Sciences* 92 (12): 5525–29.

Zimbardo, Philip. 2007. *The Lucifer Effect: Understanding How Good People Turn Evil*. Random House.

Zitzmann, M., and E. Nieschlag. 2001. "Testosterone Levels in Healthy Men and the Relation to Behavioural and Physical Characteristics: Facts and Constructs." *European Journal of Endocrinology*. 144 (3): 183–97.

Name Index

Subject Index

access to research, 5, 11
ADHD. *See* attention deficit hyperactivity disorder
adoption studies (adoptees), 21–24, 25–29, 32, 34, 47, 105, 127, 129, 162–63
aggregative analysis, 121, 136–37. *See also* nonaggregative analysis
aggression, 6, 7, 93n25, 118–19; categories of, 161–63, 173, 207; concept of, 34–35, 125, 153, 170–72, 175–77, 206–7; definition of, 35, 60, 71, 152, 154–55, 156–60, 165; measurement of, 40, 159–60, 165. *See also* antisocial personality disorder (APD)
aggression research, 3, 11, 13, 14, 153, 171, 182; dynamic systems approach, 84–86; GxExN approach, 94–95, 98; molecular behavior genetics, 52–54; neurobiological approaches, 64–65, 66–71, 72–78; public interest in, 4–5, 16; quantitative behavior genetics, 23–24, 29, 136; relevance of, 25–27, 37–39, 152, 186–87, 192–97, 199–201, 207–10;

social-environmental approaches, 41–49; uptake of, 180–92
aggressive behavior. *See* aggression; aggression research; antisocial behavior, categories of; antisocial personality disorder (APD)
AIDS, 65
aim of behavior. *See* behavior
aim of research. *See* goals of research in
alcoholism, 59, 60, 188
allelic markers, 12, 52, 54, 58, 107
allelic variation, 9, 53–54, 58, 60, 61. *See also* genetic mutations
Alzheimer's disease, 170, 171
ANOVA. *See* analysis of variance
analysis of variance (ANOVA), 21–22, 28, 29, 30–32, 36, 133–34, 141–42
animal models. *See* animal research
animal research, 55–56, 60–61, 63, 65–67, 73–78, 83, 88–92, 96–99, 107–10, 136, 140, 153–63, 167–72, 175, 182, 186n22, 190
antideterminism, 10. *See also* determinism

laboratory observation, 24, 38, 49, 84, 106, 160n23

learned behavior. *See* behavior: learned

lesbians, 11–12, 70, 74, 153, 199. *See also* homosexuality; sexual orientation

limits of research, 115–17, 203–5. *See also* criticism; scope of research

linkage analysis, 54–55, 58, 59

longitudinal research, 25, 29, 32n22, 39, 41–42, 94–95, 99–101, 105

MAOA. *See* monoamine oxidase (MAOA)

masturbation, 40

measurement, 9–14, 102, 115–17, 126–37, 145–48, 176–77, 203–6; developmental systems theory, 88, 90, 109; dynamic systems approach, 93; GxExN approach, 94–96, 98; molecular behavioral genetics, 61, 107; multifactorial path analysis, 100–101, 112–13; neurobiology, 64–69, 72–75; 78–79, 108; quantitative behavior genetics, 22–24, 27–36, 104–5; social-environmental approaches, 40–41, 45–46, 48–49, 106

measures: of aggression, 154–63, 165; of sexual orientation, 163–66; validity of 155n7, 159nn21–22, 166. *See also under* behavior: concepts of; behavior: operationalization of

mechanistic approach, 91, 144, 149

Mendelian inheritance, 59

meta-analysis, 23–24, 97–98, 162–63, 183

meta-empirical evaluation, 148

metaphysics, 9, 76, 110, 116, 138, 143–44, 146, 167

methodology, 8, 11, 132–33, 154

methods, 114–17, 126; developmental systems approaches, 81, 87–88, 92, 109–10; dynamic systems approaches 87, 89, 110; GxExN approach, 95–97, 111–12; molecular behavioral genetics, 51, 58–61, 106–7; multifactorial path analysis, 100–101, 112–13; neurobiological

approaches, 63, 72–79, 108; population-level approaches, 121; quantitative behavioral genetics, 21–22, 27–29, 32, 35, 91–92, 104–5; social-environmental approaches, 37, 44–46, 48–49, 106. *See also* analysis of variance (ANOVA); experiment; measurement

metonymy, 207

molecular behavioral genetics, 51–61, 106–7, 115; causal space of, 128; general press representation of, 188, 190; scientific media uptake of, 186

molecular biology, 143

monism, 137–38; of philosophers 141–44; of scientists 139–41, 149

monoamine oxidase (MAOA), 52–53, 54, 67, 78n40, 94–98, 111; general press representation of, 114, 188, 190, 196

mother-infant interaction, 41, 83. *See also* parents: influence of; uterine effects

motor skills, 84–85, 86, 89, 110

movement v. behavior, 93n25, 114, 166–72, 175–76

multifactorial path analysis, 99–102, 112–13, 116–17

multifactorial traits, 59. *See also* complex trait

multivariate analysis, 25, 29, 57

"murder gene," 188

muscular dystrophy, 170

National Institute for Mental Health, 38, 97

National Institute of Justice, 189n26

National Institutes of Health, 3, 53, 182

National Science Foundation, 4n5

natural experiments. *See* experiment: natural

natural kinds, 113

natural selection 140, 146, 173n48. *See also* evolutionary psychology; evolutionary theory

nature vs. nurture, 2–6, 9–12, 21, 97, 113–14, 121, 142–43, 176, 190–92, 195–201, 205–9